The CRISIS OF LIFE
ON EARTH

p. 42
51

The CRISIS OF LIFE ON EARTH

Our legacy from the second millennium

TIM RADFORD

THORSONS PUBLISHING GROUP

First published 1990

British Library Cataloguing in Publication Data

Radford, Tim
  The crisis of life on earth.
  1. Ecology. Political aspects
  I. Title
  574.5

ISBN 0-7225-2139-1

Published by Thorsons Publishers Limited, part of Thorsons Publishing Group, Wellingborough, Northamptonshire NN8 2RQ, England.

Typeset by Harper Phototypesetters Limited, Northampton.
Printed in Great Britain by Mackays of Chatham, Kent.

10 9 8 7 6 5 4 3 2 1

Contents

Acknowledgements   7
Foreword by Dr J. Leggett of Greenpeace   9
Prologue   13

Book I: Death by degrees
  1   The Great Universal Friendly Society   19
  2   The Arrhenius reckoning   27
  3   A devil in the deep blue sea   35
  4   On the beech   44
  5   Fuel's gold   53
  6   The coming of the big heat   64

Book II: The poison cloud
  7   The assault on the skies   77
  8   Lethal light and deadly air   87
  9   The hole in the stratosphere   95
 10   The dark side of the sun   107

Book III: Nature strikes back
 11   The case of the missing summer   117
 12   A bolt from the blue   125
 13   Under the ice sheets   136
 14   The message of Mars   146

Book IV: In the wasteland
 15   The great extinction   161
 16   Metropolis of root and branch   173
 17   The fruits of the forest   184
 18   A stalk on the wild side   194
 19   The kingdom of sand   201
 20   The dominion of death and life   211

Acknowledgements

This book is an attempt to tell a story for people who have never thought much about the things in it or who seek a broader understanding of them. It seemed to me that asterisks and footnotes and references would slow down the telling. For the same reason, some of the processes described may have been simplified too greatly. However, it was written by one who has pored over the Pelican library for 30 years, *Scientific American* and *New Scientist* for 20, and for almost 10 the research journals *Nature* and *Science*.

Many of the themes are distilled from learning acquired during 10 years of editing the science pages of the *Guardian*, and I am grateful to the hundreds of contributors, scientists all, whose brains have been picked for this book without their consent. Of these, the chief must be Dr Norman Myers, an environmental consultant, of Oxford, who has himself said many of these things with more authority, and more force-fulness, and long before it was fashionable to say them. He said many of them in the *Guardian*. There is a debt to climate research scientists in North America and in Britain with whom I spoke during a period as Science Correspondent for the *Guardian* — chief among these are at the Climatic Research Unit at the University of East Anglia — and also to the many conservationists, ecologists, oceanographers and environmental pressure groups such as Friends of the Earth and Greenpeace, United Nations organizations and bodies such as World Watch, the World Wide Fund for Nature and the World Resources Institute.

It is important to remember that in dealing with global processes, 'facts' are usually a matter of consensus rather than proof; and global statistics are usually a matter of informed guesswork rather than precise accounting. Even so, if there is error, it is mine. The text was read in part by Dr Nigel Williams and David Bodanis, who made many suggestions: the surviving mistakes are not theirs. There is also a debt of gratitude to the editors, and to my agent, Lucinda Culpin. And, of course, an enormous indebtedness to my wife and children, and to my late father, Keith Radford, a geographer who taught me that through books one could read the world; and that, conversely, the world could be read, too, as if it were a kind of book, as long as one studied the grammar and the vocabulary.

Foreword

Global climate change has emerged as a major scientific and political issue within a few short years. Queens now air their concern in speeches to their people on Christmas Day. Politicians shuffle for position in the wake of opinion polls showing steady escalation of public concern. Scientific journals are crammed with comment and analysis. Forty-nine Nobel-prizewinning scientists have appealed to President Bush to curb greenhouse gas emissions, professing that 'global warming has emerged as the most serious environmental threat of the 21st century . . . only by taking action now can we ensure that future generations will not be put at risk.' Even nuclear weapons laboratories host conferences on the greenhouse effect these days. Despite the predictable emergence of high-profile iconoclasts, the message is clear. Man-made greenhouse gas emissions are propelling us towards a world of unprecedented climate change. The message from the science journals is that the issue is no longer a question of whether global warming will be bad, but how bad, and by when.

In a world rapidly becoming inured to sweeping change on the political stage, we have witnessed the emergence of an environmental threat which cuts to the heart of how humans run society; a problem which is truly global in both consequence and cause. Greenhouse gases are produced in their current superabundance as a result of profligate energy consumption and intensive agriculture, in a world in which the greenhouse effect is allowed to continue its build-up, we would all — at some stage — be losers, and we would all — to varying degrees — be responsible

No quantum leap in scientific understanding explains this

situation. What we have been caught unawares by — scientists, industrialists, farmers, policymakers and environmentalists alike — is quite possibly a coincidence. The 1980s saw the emergence of new computer models which predict for the decades to come, global warming unprecedented in human history. They also saw the six hottest years in recorded history. Media-worthy manifestations of the warm 80s, whether themselves the product of greenhouse warming or natural climatic fluctuations, conspired to fix the public spotlight intensely on global warming.

The issue reached a focus in 1988. An international conference hosted by the Canadian government produced a consensus statement which spoke of effects second only to global nuclear war if humankind did not mobilize effectively and cut greenhouse gas emissions. An international body was set up to assess the full extent of the threat, and recommended responses. This, the Intergovernmental Panel on Climate Change, reported its findings to the world during 1990.

Most governments are delaying their policy decisions until the IPCC reports: many in undisguised trepidation. The implications of concerted action to cut global emissions of greenhouse gases are not for the politically faint-of-heart. Simple ameliorating measures such as investment in renewable energy production, wholesale energy efficiency, and a rethink of the economics which each year requires tens of billions of dollars to be transferred from the developing countries to the industrialized countries, are perceived by 5-year-cycle politicians and the industrial interests which lobby them so effectively as being too far outside the present frame of reference to be workable. Already, environmentalists talk of the NIMTOO syndrome: Not In My Term Of Office. Yet the first whiffs of panic are in the air. The Dutch Deputy Prime Minister, for example, told the President of Brazil in 1989 that if deforestation in the Amazon was continued to its completion, emitting as it does such vast quantities of the greenhouse-gas carbon dioxide, Holland would cease to exist as a country, flooded by rising seas.

Also taken by surprise, the relevant vested interests have now begun their inevitable spoiling tactics. The world spends up to a trillion dollars a year on its coal, oil and gas, and a further trillion dollars on its weapons. The multinational infrastructure spawned by these juggernauts over the years cannot look with relish on a world in which fossil-fuel burning must be cut to the bone, and concepts of national security trampolined from the military to the environmental.

Crisis? What crisis? Readers of an editorial in the *Wall Street Journal* on February 6th 1990 could be forgiven this age-old response. They read of the grave dangers of 'scientific faddism' and the unreliability of the 'global warming models of various agency bureaucrats.' The editorial was a response to environmentalists' criticism of a speech President Bush had given to third plenary of the IPCC in Washington a few days before. In it, Bush had spelt out that 'we all recognize that the atmosphere is changing in unexpected and unprecedented ways', and 'we know the future of the Earth must not be compromised.' Fine sentiments, but — to the exasperation of many present — the President made no specific commitments to buy insurance against such eventualities. 'The politics and opinion', the President informed us, 'have outpaced the science.' Sitting in the audience, I struggled to assimilate this Orwellian reversal of what I see each week in the scientific journals. 'Wherever possible,' the President continued, 'we believe that market mechanisms should be applied and that our policies must be consistent with economic growth and free market policies in all countries.' Heaven help us if corporate interests are unleashed unfettered on this problem. Does the President believe General Motors is going to roll over and do the right thing without substantial federal inducement? The *Wall Street Journal's* editorial writer spoke for corporate America a few days later: 'We hope the President hangs tough on this one.'

Readers of Tim Radford's book on the crisis for life on Earth might be forgiven for wondering whether humankind is indeed capable of fashioning itself an escape route. All today's decision makers will be dead and gone when the worst impacts of global warming, as predicted by the current climate models, arrive. This may even be true if the very worst-case analysis comes to pass: a coincidence of positive feedbacks which causes a runaway greenhouse effect setting us on course for an atmosphere of the kind favoured by planets devoid of life. Progress thus far has indeed been glacially slow. The IPCC may yet prove fatally compromised by the intransigence of the big greenhouse gas emitters, led by the United States. Rhetoric has often been impressive, progress always minimal. We hear talk of the need for stewardship and see cuts in budgets for energy efficiency. We see leaders queuing to host international seminars, yet not committing to even freeze — much less cut — carbon dioxide emissions. The IPCC scientists' working group professes that a cut of carbon dioxide emissions of up to 70 per cent will be needed to stabilize the level of this particular greenhouse gas in the atmosphere.

Yet progressive cases have formed. The Dutch government, and the state governments of Vermont, USA, and Victoria, Australia, have committed themselves to carbon dioxide cuts by the end of the century. So has the city of Toronto. And we live in a world where paradigms are shifting, walls collapsing, week by week. In the few days I have spent borrowing spare moments to put these thoughts on paper, I have seen the Central Committee of the Soviet Communist Party vote for a multi-party state, and seen Nelson Mandela walk to freedom. One thing is certain, we will be living in interesting times in the 1990s.

Dr Jeremy Leggett FGS
Director of Science
Greenpeace UK

Prologue

On the evidence of the *Wall Street Journal* of 27 June 1988, millennial paranoia has already struck. Mr Richard Kieninger of Adelphi, Texas is racing to assemble a fleet of blimps, dirigibles or airships to form an ark, in which a few hundred people can float serenely on a cushion of air on 5 May, 2000 AD, and at the same time watch the destruction of the world immediately below them. Mr Kieninger himself is quoted as saying that 'It's kind of far out, but it's wise to prepare. Some people could be driven mad just due to the terror of it.' The destruction he has in mind will be a series of terrible earthquakes, tidal waves and volcanoes, triggered by a massive buildup of ice at the South Pole which will force a catastrophic shift in the earth's axis. This will (he thinks) happen because Mercury, Mars, Jupiter, Saturn, the Earth and the Sun will all be in alignment on 5 May, 2000 AD. It's the kind of belief that ripples through the most sensible communities at the end of an era: at the end of each century tiny madnesses spring up; and at the end of the last millennium whole Christian sects (the only ones, of course, who counted the years to 1000 AD) awaited the Second Coming, with the attendant Four Horsemen of the Apocalypse. There is a name for this condition: it is called millennarianism.

Mr Kieninger is not alone in his anticipation of imminent disaster. The last two decades have been marked by evidence of a widespread unease about the human race and the world about us. The irony is that this unease is occurring at a time when the expectation of health and longevity, not just for the rich nations but also for the poorest, is probably higher than it has ever been, and this blessing may now be conferred upon more people than have ever existed in the whole history of the globe; when health, services and food supplies and hygiene regimes have advanced everywhere; when education and communications technology have spread over the whole globe; when science and engineering have conferred, even in the past few decades, an astonishing power upon humanity; most of it a power for good, for wisdom, for maturity, for individual happiness.

We have become a little casual about this power, and take some of it for granted. We are beginning to understand the mechanisms of life down to the smallest detail: we are preparing to map the entire genetic structure of the human species; we have built vast particle-smashing machines which are helping us assemble the ultimate fabric of matter; we have cosmologists who can speak confidently of the nature of the universe millionths of a second after its creation; we have devised machines which can punch through the Earth's gravitational hold at a speed of six miles a second, which have explored the solar system and which are already heading into deep space, and we carry in our pockets tiny calculators which can, in a few seconds, calculate to powers of 10^{99}: that is, 10 multiplied by 10 by 10 and so on 99 times. This is a number far greater than all the atomic particles in the universe, and we don't give it a thought. We spend more time thinking about 'designer' clothes than about 'designer' animals and plants, but we have begun to explore the possibility of fashioning creation to our needs, not by the old crossbreeding methods of trial and error, but by precision work in the genetics laboratory.

We have even begun to question death: not seriously, because no one so far has cheated it, but we have extended life's domain by medical technology so far that the definition of death has become a philosophical and legal point, and there are already commercial enterprises devoted to preserving human bodies in the hope that they may be revived. Most of this knowledge has been bought relatively cheaply — it rests on human imagination, a capacity for sustained logic and a tradition of experience, observation, and experiment, but it also rests on a series of strata of economic growth and expansion that, in the last 200 years, has changed our expectations utterly.

But even so, Mr Kieninger is right to be worried. His unease is now shared by climate scientists, economists and governments everywhere. The irony is that there is a reckoning in the years ahead, and that it will probably begin to be felt everywhere by the turn of this century, which will mark the end of the second millennium. It will be felt because, for the first time since human beings occupied the planet, our activities are powerful enough to alter the balance of its atmosphere. And by doing this we have also altered the distribution of the water and the structure of the soil on which our future depends.

The Hebrew Book of Genesis saw it coming: 'And God blessed them, and God said unto them, Be fruitful and multiply and replenish the earth and subdue it, and have dominion over the fish of the sea, and over the fowl of the air, and over everything that moveth upon the earth.' What frightens anyone who thinks about it is that this seems to have become almost entirely true. For the first time since creation the survival of the wild things — at least the larger ones — is entirely in our hands. The wilderness is at hazard: at the present rate it could *all* be tamed or subdued within the lifetime of a child born today. The activities of the industrial nations of the West and the Eastern power blocs have altered the composition of the earth's atmosphere, not just locally, not only around cities and industrial towns, but everywhere. These same activities have begun to warm the globe: within a few decades the world could be warmer than it has been for 100,000 years, directly because of the burning of coal, oil and natural gas. Industrial developments have begun to threaten the composition of the ozone layer that protects us all from the fiercest rays of the sun. Huge tracts of the world are turning to desert: the seas and rivers are now being exploited so hard that fish supplies are beginning to dwindle. The spawning beds, that might replenish the catch, are also being polluted. At the same time human ingenuity is extending the lifetime of all those living today, and offering an ever greater chance of survival for the newly born, but then that, too, is part of the problem. The same ingenuity offers a chance of resolving the problems that will follow in the train of the greenhouse effect and the destruction of the ozone layer; of halting the waste of the forests and the spread of the deserts. These things are not the only problems facing humanity, but they are the biggest, and they are problems for humanity as a whole, rather than segments of it.

This book is an attempt to explain how and why these things are happening: a diagnosis, but not a cure. A problem like the greenhouse effect, or for that matter species extinction, will not be solved by any single solution. It may not be solved at all. We all may simply have to adjust to a very different world. But this is not meant to be a gloomy narrative: it is simply another way of looking at geography, which is a dull word. But the world is not dull. It is also a book about the future. People who find its conclusions depressing are offered this cheerful thought: that there have been, over the last century or so, many books about the future and most of them, as far as we can judge, have turned out, to a greater or lesser extent, to be wrong: often right in all sorts of minor details but grossly wrong on one important head: or broadly right but wrong in failing to ancitipate how we will react. With luck, too, this book will be wrong. But it is not based on the conclusions of one man: it is based on the findings of laboratories, government agencies, international panels, environmental organizations and the scientific advisers of political leaders. And there is another thing to fear: if this book is wrong it may be that it is false prophecy only in that it does not paint the future in dark enough colours. Richard Kieninger, may, if he soars high above it all on 5 May, 2000 AD, be doing the sensible thing. Nature has her own way of responding to disturbances of her; nor does she even need to be disturbed to strike at us. Plague, bloodshed and famine may be things we bring upon ourselves by our own actions or inactions, but earthquakes and volcanoes and Ice Ages are not of our prompting, or taming. But Mr Kieninger in his fleet of blimps high above the new-born 21st century won't have escaped the future, because it is in the air above our heads that the troubles have started. And with him, in the air, is a good place to begin our story.

BOOK I

Death by degrees

CHAPTER 1

The Great Universal Friendly Society

We carry the burden of our future and our past upon our shoulders and we don't feel a thing. We talk about being as light or as free as the air and we don't give it another thought.

It isn't light. The heady mixture of oxygen, nitrogen, argon, carbon dioxide, neon, helium and methane and a few traces of other gases presses down upon the earth with a weight of 5000 million million tons. We don't pay any attention to it because it's just right for us, and we are just right for it. We are at home in air. We couldn't live without it. Nor could a mouse, a cockroach, an elephant, an anemone, a whale, an olive tree or a blade of wheat. The internal pressure that holds us together is in perfect balance with the colossal tonnage that bears down upon us. It is the medium in which we evolve, grow, climb, fly, stand, love, die and decompose. Without it, our smallest cells could not divide. Without it, our bodies could not rot. It is our medium: from ash seedling to ashes, from feather duster to dust. We don't pay attention to it. And if it wasn't there, we wouldn't notice its absence; we, too, and everything we recognize as alive and many things we don't, would also be absent. The air is the blank page upon which all life writes itself, and is after a while erased.

It isn't free. It is a bank upon which we draw, and into which we make a payment, every time we breathe, plant a lettuce, throw the lawn clippings onto the compost heap, start the car, fell an elm, burn the apple logs, bulldoze a forest, smelt iron ore, harvest cotton, weave cloth, order a shirt, incinerate poisonous waste, mow a meadow or toss hay to the cows. It doesn't help to think of it as free. It helps to think of it as a universal circulating currency, managed impartially and without judgement in fear or favour by a global, sky high

Friendly Society: a society of which every living thing is a member, all debtors and creditors at the same time; a society which up till now has needed no managers, no audit of accounts, no boardroom strategies, no research department, no investment planning, no policies on rates of exchange, no fixing of interest, no calling in of loans, no foreclosure.

The notion of the air as a bank or a Friendly Society account for continuous withdrawal on demand is not an empty conceit: it is, strictly, Our Bank. We, that is Life, Nature, Evolution, the biota — or, to put it more vividly, the protozoa and the gymnosperms, Tyrannosaurus rex and the coelacanth, the grasses and the liverworts, the tundra and the rainforests, the ungulates of the African Rift, Homer, Cleopatra, the Knights Templar of the Crusades, Shakespeare, Lucretia Borgia, Jack the Ripper, Marie Curie and Albert Einstein, the corals of the Red Sea and the tomatoes of the Andes, the volcanoes of the Pacific, the rolling chalk Downs of Sussex and the jagged limestones of China and Yugoslavia — have made the bank what it is today. Life itself — and the very rocks are part of it, and not just those like the White Cliffs of Dover or the tar sands of North America that were once sentient, growing things in the primeval ocean in the morning of the world, but also the deep, the seawater itself, and the muddy sediments and the incandescent pyroclasts flung upwards from the craters of volcanoes, that we think of as dead, inert, never having lived — paid in, and withdrew, and in doing so altered the nature of the fund, subtly and imperceptibly adjusting its composition so that the nature of currency is just exactly what we need for all of us to be what we are, right now, with more than enough to go round.

We don't know what the Earth's original atmosphere was made of, more than 4500 million years ago when the planets of the solar system first formed. Our best guess is that it was composed of carbon dioxide, nitrogen, water vapour, methane, ammonia. Atmosphere — any atmosphere — is the gift of the molten rocks in the planet's crust and mantle. 'As if this earth in fast thick pants were breathing', wrote Coleridge in *Kubla Khan*. There is no 'if'. The rocks have always breathed, sometimes in susurrations: slow, steady, rhythmic upwellings of magma; sometimes in explosive eruptions that have darkened the skies and seared the landscape and shaken the seas. We can't be sure about the earliest volcanoes, but the craters of Hawaii today exhale water, carbon dioxide, sulphur dioxide, nitrogen, hydrogen, carbon monoxide, chlorine and argon. Water makes up 77 per cent of the volume, carbon dioxide about 11 per cent. But the world's first air was not called upon just to support the world's first life, but to be both cauldron and ingredient in its creation. There must also have been methane and ammonia. These, and the other gases, fired by the sun's ultraviolet rays or lightning or both, provided the first organic chemical stew, in which was cooked up the complete molecules called amino acids and nucleotides: the broth in which the stock of all living things was formed. This is not just conjecture: the recipe was kitchen-tested in an American laboratory in 1953 by Stanley Miller. In a week, using pure, sterilized water, hydrogen, methane and ammonia and an electrical discharge, he made a very dilute soup containing two of the amino acids found in living tissue. What took Stanley Miller seven days, the planet could easily surpass in a few billion years.

But no such things happened on other planets. On most of them, conditions were impossible: too far from the sun, or too near. It is not impossible that life might have started on Mars, but if it did, it came swiftly to a dead end. Life might once have been possible on Venus, but we now have no way of knowing. But the launch of life is in the realm of conjecture: there have been arguments that the raw material for life could have been seeded by the arrival of meteorites, some of which are rich in organic compounds. We don't know.

The one thing of which scientists are certain is that the Earth's first atmosphere had virtually no oxygen. The raw material of life cannot survive in an oxidizing atmosphere. For a billion years or so what passed for life must have been something most of us would not recognize as life at all. But about three billion years ago, tiny single-celled creatures developed a new life agent: photosynthesis. They used the light from the sun, carbon dioxide and water to build more cells, and excreted, as a by-product, oxygen. They couldn't live with the oxygen, of course: they produced it as an iron oxide, or rust, and they left it behind in the geological formations which girdle the earth, and everywhere date an evolutionary cul-de-sac. But new creatures, that could live with their own oxygen waste, began to flourish. Within a billion years the very existence of a proportion of oxygen in the atmosphere began to halt the progress of evolution in one direction, and direct the traffic of creation in another.

The story is richer and wider than one of life and atmosphere. The new creatures that colonized the hot bare rocks and the oceans began to lower the temperature of the globe. When the first life forms began about 3.8 billion years ago, the oceans may have been at what is now boiling point — 100°C. This is not a guess: there are bacteria today in thermal springs and volcanic magma, deep in the fissures of the earth's crust and under the seas at the point where hot mineral brines and basalt magmas surge onto the ocean floor, at great pressure. They do best, that is they thrive and multiply, at up to 105°C. The life that flowed from such creatures and their predecessors also speeded up the weathering of the rocks. The soil that is creation's bedrock, so to speak, is also a creation of life itself. One estimate is that if there had been no life on earth, the temperatures would be about 45°C higher than they are now. Each enrichment helped to step up the rate of weathering, produce a little more oxygen, reduce the proportion of heat-holding carbon dioxide by a fraction, and open the way for new creatures to hold their place in the sun. Some of the earlier creation survived: anaerobic bacteria — organisms which live and work without oxygen — still play a vital part in making the world work. But even so, quite literally, the green plants had switched on the green light for a green world.

Two billion years ago, oxygen and nitrogen began to become major components of the atmosphere, and carbon dioxide a minor one. We — footballers and speckled thrushes, hairdressers and centipedes, vacuum cleaner sales reps and wildebeeste — did not become inevitable by this means. Evolution has its routes in chance and necessity. Individuals and species and genera are only the by-blows of chance. The necessity was that, in whatever way life evolved, it would end up in a balance of carbon-based life-forms using photosynthesis to build themselves with the carbon in carbon dioxide, and creatures using the oxygen surplus to digest and recycle the carbon and return some carbon dioxide to the air, and the rest to bones and stones. To skip a couple of billion years, by the beginning of the Industrial Revolution, and at precisely the time that poets, painters and musicians began falling in love with the idea of 'Nature', the 5000 million million tons of the earth's atmosphere had a composition of 99.97 per cent nitrogen, oxygen and the inert, or noble, gas argon. Carbon dioxide, without which deep sea divers, economists and football hooligans could not exist, nor have come into existence, had dwindled to little more than a trace gas: just a couple of thousand billion tons altogether, or 270 parts per million by volume of the atmosphere as a whole. The importance of that tiny proportion — to understand how tiny just think of a slice of lemon in a bathtub full of gin and tonic — cannot be understated. The carbon dioxide proportion in the last 200 years has risen to just 330 parts per million. Same gin and tonic, same bathtub and the twist of lemon is now just under a millimetre thicker, and yet climatic scientists, politicians and environmentalists are beginning to talk as though the end of civilization as we know it is at hand.

It may very well be at hand for some people. This is because the greenhouse effect — the name was coined at the end of the last century — is a natural, benign agency that may be adjusting just a shade too fast for us. It is not like other crises of atmospheric pollution: not like the hole in the ozone layer or acid rain or photochemical smog or toxic metal pollution. All of these have occurred because we have been releasing into the atmosphere in large quantities, things which were not already there except in traces. All of them have, until recently, been regarded as a necessary or acceptable price to pay for the creation of wealth and comfort. We have paid for them with our health and our amenity and also in large quantities of cash (some years ago acid rain was calculated to cost Europe alone up to £44 billion a year in damage to agriculture, forestry, buildings and health) but the industrial wealth we have bought with them has also, in the Western World, enabled us to live on balance healthier and richer lives.

With that wealth in the bank, the richer countries have already started to tackle the problems of acid rain, smog and stratospheric ozone pollution. They know, on the whole, what must be done, and how to do it, and — although some are still dragging their feet — are beginning to do it. The problems are (except in the case of the ozone layer) local or regional; the solutions rest in finding ways of not pumping sulphur dioxide, or arsenic, or carbon monoxide, or lead, or chlorofluorocarbons into the atmosphere. The message is the same. Don't do this (or do that) and the problem will quickly (or eventually) clear itself up. The balance of nature will be restored.

But the greenhouse effect *is* the balance of nature. All that has happened is a tiny, sudden bull market in the circulation of carbon dioxide in the great atmospheric Friendly Society finances. It has happened in the past; the rocks and the ocean ooze and the ice cores of Greenland and Antarctica reveal a record of intermittent hiccups — slight imbalances in the deposits and withdrawals or carbon dioxide — in the accounting long before human history. The causes may have sometimes been dramatic, but they have been completely natural. Even the present crisis is, ultimately, natural. The carbon dioxide being released from fossil fuels burned in power stations and motor vehicles is still part of the atmospheric bank: it has merely been held in reserve for a while before being returned to circulation. If the same slight increase in parts per million by volume were to happen over 500 or 1000 years neither humanity nor the plant nor the animal kingdoms would notice it: the adjustments would be infinitesimal and imperceptible. What fear can there be of seas that rise a millimetre in a few decades? Or temperatures that adjust by a fraction of a degree a century?

To us the air is not just light, it is translucent. The radiance of the sun hammers the land and the sea, the deserts, forests, ice-caps, savannahs and cities with a power to raise 1.4 kilowatts of energy per square metre: if you could use it you could run a hair dryer forever. But most of it bounces back into space immediately, or is scattered in the atmosphere, or is released later in the form of infra-red radiation. If the atmosphere were composed only of oxygen and nitrogen, the sun's magnificence would radiate straight back through the translucent medium of the air. But a few trace gases which are as glass to visible light are in fact opaque in the infra-red: they absorb it. We can't see this. Some reptiles, as it happens, can. Their skies are different. They are tinged with brown. Pit vipers, and rattlesnakes can actually see the greenhouse effect. They can see what we can only feel: the warm second skin that keeps the world at room temperature.

Without the greenhouse gases, the world would be a bleak, inhospitable place. If you could vacuum from the skies every trace of carbon dioxide, methane, nitrous oxide and water vapour, the global average temperature would plummet by about 33 °C. Believers in the story of Genesis may justly point to the happy accidents of warmth, water and atmosphere (enough oxygen for us to breathe but not enough to make fire a permanent hazard and spontaneous combustion a daily event; water which expands as it freezes, so that it floats and insulates the liquid water below it; an atmosphere which has just enough greenhouse gases to keep the globe at a habitable temperature for four-fifths of its land mass) and see within all this the hand of a caring Providence. Darwinists point out that nothing else could be expected: we evolved with the atmosphere. Of course it is just right for us.

In fact, the levels of carbon dioxide must have fluctuated dramatically throughout geological time: even in recorded history the evidence tells of bouts of uncharacteristic warmth and inhospitable cold. Medieval monks harvested and trod grapes in Yorkshire; seventeenth century Londoners roasted an ox on the frozen Thames. But the shifts in average temperature need only have been very slight. A fall of 1°C could take Britain back into a Little Ice Age. A fall of 4-5°C would bring the polar glaciers back into southern England. Nobody knows what a rise of even a couple of degrees would mean for the world. Even a one degree rise would bring the globe to a temperature higher than any for the last 100,000 years.

We know something about the patterns of climate in the past; we know about the climate we've got; we cannot know about the climate we might have, because a single question about climate means hundreds of questions about the seas, the deserts, the forests, the tundra, the ice-caps, the mountains and the volcanoes; about the ocean currents and the mixing of air masses all the way up to the stratosphere; it means questions, too, about the behaviour of the sun, the rhythms of the solar system and even the progress of the sun round the galaxy. It also means questions about how plants behave, about what happens when we switch from one way of using land to another.

Above all, it will mean questions about people; about political decisions, about population growth, about development, about energy use, about agriculture and famine. What happens in a greenhouse world will depend on the decisions which all nations take: these decisions will *have* to be taken on the basis of an understanding we don't yet have; and some of them will *have* to be taken on the basis of little more than guesswork, or the impact of the greenhouse warming will be even worse. The final headache is that the decisions aren't all that obvious. There isn't any one action that any one nation can take that will solve or even much alleviate the problem. There will have to be hundreds of courses of action, taken in many cases in concert, and involving a surrender of exactly the ambition and sovereignty which have underpinned the idea of nationhood since history began. The notion that individual nations should take decisions that may cost them wealth and power for the common good of all humanity is one that many will subscribe to in principle, in the hope of being revered for their farsightedness. In practice hardly any of them have ever acted upon such principle. Anyone who feels foreboding about the world in a greenhouse warming has history on their side. These are not cheerful days for optimism.

CHAPTER 2

The Arrhenius reckoning

The Swedish chemist Svante August Arrhenius was one of those people who spent his life ahead of his time. He taught himself to read in 1862: he was 3 years old. His doctoral thesis at the University of Uppsala put forward the theory of ionic dissociation to explain what happens to salts during electrolysis: his examiners were incredulous, and reluctantly awarded him the lowest possible passing grade after a four hour examination. The same dissertation earned him the Nobel Prize for Chemistry in 1903.

Some of his other ideas are still a matter of controversy: he proposed the notion of 'panspermia' in 1908: that is, that living spores were driven from star to star by radiation pressure, and that the seeds of life being everywhere, life need not have spontaneously begun on Earth. Most people don't care much for this idea, not because it is impossible or even improbable — astronomers using spectroscopic analysis, a technique which 'reads' the light passing through molecules, have found organic compounds in the space between stars; they have found infinitesimally small diamonds and organic matter in meteorites that can only have come from Mars — but because it simply shifts the question from the zone of the answerable to the unanswerable. That is, if we accept that life began on earth, we can by experiment, analysis and observation, at least try to work out how. If we believe that it was blown to us by chance from a distant star, then we can't.

At the close of the last century, Arrhenius turned his attention to the matter that now exercises the world's scientists and political leaders: the radiative properties of gases. That means how gases absorb heat, and release it again, in the way that a household radiator absorbs energy in the form of electric current and releases it in the form of heat. He suggested that carbon dioxide in the atmosphere formed a heat trap. High frequency sunlight could sail through the air to warm the earth and water. The low frequency infra-red heat re-radiated back at night would be trapped and held by the atmospheric carbon dioxide. It would take, he suggested, only a slight rise in carbon dioxide to raise the world's temperature high enough to account for the clement climate of the Mesozoic Era. This is the era of the dinosaurs. A slight fall in the levels of carbon dioxide, the argument went on, might explain the Ice Age. 'Such a suggestion,' remarked Isaac Asimov in 1964, 'is still taken seriously and, indeed, may account for the situation on the planet Venus.' The planet Venus has a surface temperature that would melt lead. It is screened by clouds of sizzling sulphuric acid, and its atmosphere is mostly carbon dioxide. It is a prime example of the runaway greenhouse effect.

Since, on the whole, humanity has settled down quite comfortably with the greenhouse effect of the last 100 years or so, it should be quite clear that current concern is really about the speed of change. Sometime during the next century the amount of carbon dioxide in the atmosphere is expected to double. The predictions have settled on a probable increase in global temperatures of somewhere between 1.5°C and 4.5°C before the year 2050 AD; some of the predictions put an earlier deadline of 2030. The increase is not likely to be uniform: that is, 1 January 2030 will not be 1.5 or 4.5°C hotter in Hayward's Heath, Sussex or East Cicero, Illinois than it was in the same place on New Year's Day 1990, nor will there be a similar rise in either place on 30 July. Climate doesn't work like that, nor, as everybody must have observed by now, does weather. Winters will be warmer, but will not necessarily be matched by hotter summer days. On the other hand, those areas subjected to droughts and heat waves in the summers will see more of them. Even though climate scientists and environmentalists keep pointing out that most questions about a greenhouse world are, right now, not answerable, there is a consensus of sorts: that for instance, there will be a greater frequency of 'extreme events' — droughts, floods, storms.

There is another consensus. Just as the rising temperatures will 'flatten out' in the summer, so too will the big heat spread its blanket unevenly over the latitudes. The equator will get a little warmer. Some temperate regions may not notice much difference. But the high latitudes, the sub-Arctic and Antarctic, will get a lot warmer — up to 12°C. The ice-caps will retreat, the glaciers begin to melt. Sea levels will begin to rise.

They will rise for two reasons. One is simple. Heat water and it will expand a little. Heat an ocean and it has nowhere to go but up. The extra half a degree that the world has recorded this century — not necessarily because of greenhouse warming, because there have always been slight oscillations in world temperature and this may simply be another — may have already forced part of a worldwide sea level rise of about 10 centimetres. It's difficult to be sure. It isn't easy to measure sea level accurately. The water doesn't stay still; the tides rise and fall; the high and low tide levels themselves shift according to a cycle; there are storm surges; and (it became clear from measurements taken by Seasat, a shortlived ocean satellite) the sea itself is marked by hills and valleys according to gravity anomalies in the rocks beneath. And then there are statistical problems. How long have people been keeping records? How accurate were they in the first place? And finally, of course, the land — the only base point from which you can measure sea level — is sinking. Or rising. Or being reclaimed, or built upon, or eroded, or bulldozed. But even so, it has become clear that not only are the oceans getting warmer; they are rising. Some of the rise is certainly attributable to thermal expansion. The other reason must be melting ice.

Most of the world's fresh water is actually locked up in the ice-caps of Antarctica and Greenland: if it started to melt at a rate greater than that of its formation, the world would be in trouble, because most of the world lives near sea level, and would be inundated. There are fears that this may be happening. They may have no foundation — yet. Ice-shelves break up, drift away and dwindle in the polar summers, and reform in a slightly different pattern every winter. Glaciers calve, and their silent, ghostly infants, the icebergs drift towards warmer seas not according to average global temperatures but according to rates of flow and the mass of ice building up behind them. Once again, the pattern is slightly different every year. Once again, it is difficult to measure, and difficult to be sure of the accuracy of historic records, such as they are. But some melting of the glaciers is certain.

In 1989, the UK's popular press flooded, so to speak, the popular imagination with predictions of a 5 metre sea level rise, and a drowning of the Houses of Parliament on the banks of London's Thames. The more sober, and undoubtedly more accurate consensus (because it would take a very long time to melt that quantity of ice) puts the rise in sea levels by 2050 AD (or 2030) at between 20 centimetres and 1.65 metres. And the British savants of the greenhouse effect — the Climatic Research Unit at the University of East Anglia at Norwich, led by Professor Tom Wigley — expect the rises in both temperature and sea levels to be nearer the low end of the scale than the high one.

But that doesn't mean that within those rises there won't be the makings of a world-wide disaster. There are several reasons why nobody should feel complacent simply because parliamentarians will almost certainly have both their heads and their feet above water in 2030 AD. One is that because of inertia (when you light the gas under the kettle it takes a few moments for the water to begin to get hot) the lag between the greenhouse cause and the warming effect is put at about 30 years. That is, if the temperature is up 1.5°C and the sea levels up to 30 centimetres in 2030 then both are certain to rise a great deal higher in the next quarter of a century. Another is that even an apparently slight global average temperature shift could have a massive distorting effect on climate patterns, regional weather and national agriculture. And even a small rise in sea levels could prove expensive for many nations and, for a few, catastrophic. There is a third factor to be considered: carbon dioxide is not the only greenhouse gas. Nitrous oxide, ozone, methane and water vapour and some strictly unnatural, that is, artificially produced, gases such as CFCs also play a heat-trapping role. Most of them are better heat traps than carbon dioxide. All of them are adjusting in the atmosphere for different reasons at different rates and by different processes.

The fourth factor is even scarier. It is that the greenhouse effect may be a process that continually reinforces itself. That is, it might be a dynamo that uses its own output to accelerate itself further. Engineers know the phenomenon well. They call it positive feedback.

The other greenhouse gases

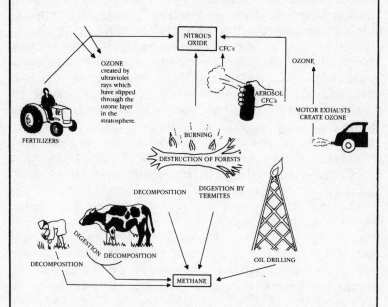

About half of the global warming expected from greenhouse gases can be attributed to carbon dioxide, which is only 0.03 per cent of the atmosphere by volume. It is expected to double in the next 50 years from fossil fuel burning and deforestation.

About 18 per cent of global warming will be from methane, or marsh or natural gas: this is released by industry, agriculture, animal husbandry and the termites that feed on the destroyed forest timber. Concentrations are increasing.

Nitrons oxides from fertilizers, traffic and industry make up about 6 per cent of the problem, and CFCs could account for 14 per cent. Low level ozone could account for about 12 per cent.

This is because the atmosphere — to return to metaphor — is more than just a bank or Friendly Society. It is a global currency and commodity exchange for all living things and all the apparently lifeless phenomena upon which our lives are based; based literally, for we are ourselves but mostly water, carbon, nitrogen from the air and a few trace elements from the soil; based literally because all life makes its brief stand upon a stage of rock and water. The great, complex, interlocking currency and commodity exchanges make no exceptions for individual life forms; they work because everybody withdraws and everybody pays back and although some of the commodities are banked for a while — wood, coal, oil, peat, hay, permafrost, bone, coral or limestone for instance — they are only part of a reserve which the living world can call upon again. The only new feedstock, the only new or outside investment upon which the members can capitalize is, fortunately, as regular as clockwork. It checks in daily in each part of the globe, at a strictly predictable time to open a new round of trading.

The engine that drives the myriad dealings of the living world is the heat and light of the sun. The markets which open with the sun's first rays, and go on trading for a night after it moves on have for scientific convenience been given names and then mapped. They are called things like the energy cycle, the hydrological cycle, the nitrogen cycle, the carbon cycle, the oxygen cycle and so on. They are more subtle, and more open, than say the Stock Exchange or the Metal Exchange or the currency markets in a great capital city such as London or New York or Tokyo. Most living things are members of all of them simultaneously, and were any one cycle to shut down, the crash that would follow instantly would be terrible. But they don't shut down. They may falter in one area or another because of occasional cataclysm; whole groups of investors have checked out quite suddenly and withdrawn into oblivion, leaving only a few fossil traces in the cliffs of Lyme Regis in England, or in Texas, or the Gobi Desert. At intervals, and for some families of living things, the global bank and exchange has turned into a casino, but each loser's loss has been some other winner's gain. The territorial sovereignty that the reptiles surrendered at the end of the age of the dinosaurs, was inherited and modified by the warm-blooded creatures that suckled their young.

The processes are so complex and interlocking that a British atmospheric chemist, Professor James Lovelock, captured the imaginations of many environmentalists and scientists by proposing the world as a living entity itself: a Life Force, or Earth Mother that kept itself in continuous renewal and repair. He called his hypothesis Gaia. Most scientists use a duller name for the sphere-shaped skin, sky high and ocean deep, in which all life resides and trades and replicates. They call it the biosphere.

The cycles of the biosphere are easier to understand than to explain; that's because we know them, and rely on them, without needing to comprehend. The sun's power slams onto earth, air and water. Plants use it as a power source to take up the nutrients of soil and water and air — carbon dioxide, nitrogen, minerals and so on — to make flowers, fruit, leaves and wood. They respire some of the energy as heat. Some of the remaining energy is consumed as fodder or forage by herbivores, in turn eaten by carnivores and some of it goes in decay. Think of it in terms of lettuce, rabbit, gardener and compost heap. The lettuce grows; a rabbit consumes it. Exasperated at the destruction, the gardener hurls the stump and the half-eaten leaves on the compost heap, shoots the rabbit and eats it instead. He is, ultimately, eating the energy of the sun. So did the rabbit. So do the microbes on the compost heap: put your hand on a deep pile of lawn clippings, potato peelings and empty grapefruit, and feel the energy dissipating in decay. For a while, the gardener himself stores some of the energy. He too dissipates some of it in the exercise of breathing or gardening but during his lifetime he represents a 70-odd-year store of solar energy. When his turn comes to die and decompose the last nutrients of all the lettuces and rabbits he has eaten will be returned to be used once again by some other creature in the biosphere, for the nutrients have all of them been but borrowed. The only free lunch in the biosphere is the heat of the sun.

There is a possible variant on this theme: gardener, rabbit and lettuce could have perished, side by side or one within the other, in conditions which would make complete decomposition temporarily impossible. They could have been flooded in low-lying land and buried in wet mud. They could have been preserved in peat (the lettuce could have become peat) and eventually sealed in frozen soil —

permafrost — or they could have partially decomposed into
the constituents of fossil fuel: one of these is sapropel, the raw
material of crude oil. But the return to the biosphere has only
been delayed for 100 thousand or 100 million years or so.
They would have joined the reserve bank, or more properly
two reserve banks: one of carbon, the other of energy. They
would remain reserve units of currency in the biosphere,
ready to be cashed one day in the atmospheric bank and
traded once again on the exchange in the sky.

But there are other vital solar-powered processes going on
at the same time. One of them is the water cycle. The engine
of the sun evaporates water from the land and the oceans and
returns it as rain; it may fall and evaporate again, or it may
become temporarily stored in a lake, or as groundwater
trickling through fissures in the rocks to reappear a few billion
years later in a Volvic or Perrier bottle, or trapped for a short
lifetime in the tissues of grass, or for a long one in the frozen
bogs of the tundra, the glaciers of the Rhone or the polar ice-
caps. Water is a remarkable medium. One of its most
remarkable properties is that as it gets colder it becomes
denser, but as it nears freezing point it expands. Another
remarkable property is that it can exist in three states at the
same temperature: as water vapour sublimating on the surface
of the polar ice, as ice floating with 90 per cent of itself
submerged in water that hasn't frozen, and won't because it
can keep moving under its own insulating shield of ice.
Further, both water and ice can conduct light. Further still,
water is a superb conductor of heat: no liquid can match it
except mercury. It can also store heat better than any other
fluid. Another extraordinary property is that it can dissolve at
least small amounts of almost anything. One of them is
oxygen. Another is carbon dioxide. And it is the role of the
oceans, which cover 70 per cent of the globe, which make the
greenhouse effect something to fear, and yet something to
marvel at.

CHAPTER 3

A devil in the deep blue sea

The world is, above all else, wet. The oceans cover 70 per cent of the globe. The biosphere contains about 1500 million cubic kilometres of water, above and below ground, in the glaciers, rivers, ice-caps and oceans. About 97 per cent of it is in the oceans. Most of the remaining 3 per cent is in the ice-caps. The percentage in the atmosphere at any time is tiny. If you modelled the earth into a perfect sphere, with no bumps, for the continents on it, and then arranged the water in layers, the fresh water of the lakes and rivers would cover less than one metre. The water normally found underground would cover anything from 15 to 45 metres, the ice and snow about 100 metres, the oceans around 2700 metres. Not surprisingly, therefore, at any time they contain about 95 per cent of all the world's 'mobile' carbon, that is, the carbon that is extracted from carbon dioxide by plants, goes to build the bodies of animals that eat it, and is eventually returned to the air by burning or decomposition.

If you say the figures quickly, you can remove them of meaning: the oceans each year soak up from the atmosphere 105 billion tons of carbon and return 102 billion tons. Humans burn between 6 and 8 billion tons of carbon every year, either by setting a torch to the forests, or igniting fossil fuels, but only about 3 billion tons stays in the atmosphere to add to the greenhouse warming. To put it another way, carbon dioxide in the atmosphere is increasing each year at a rate of 1.5 parts per million instead of 3 ppm or more. Although some of the extra carbon must be feeding terrestrial plants, the only conclusion to be drawn is that the ocean is somehow taking up perhaps a third, perhaps half the surplus. Furthermore, it seems to have taken up about half of all the extra carbon

The carbon cycle

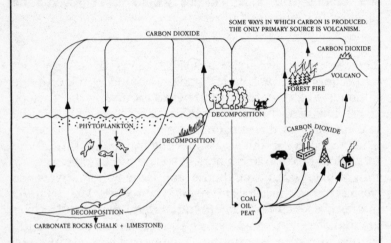

SOME WAYS IN WHICH CARBON IS PRODUCED.
THE ONLY PRIMARY SOURCE IS VOLCANISM.

CARBON DIOXIDE

CARBON DIOXIDE

VOLCANO

FOREST FIRE

DECOMPOSITION

PHYTOPLANKTON

DECOMPOSITION

CARBON DIOXIDE

DECOMPOSITION

COAL
OIL
PEAT

CARBONATE ROCKS (CHALK + LIMESTONE)

The earth's atmosphere (like Mars and Venus) was probably once mostly carbon dioxide. Green plants on land and in the oceans have been 'fixing' the carbon in it as fossile rocks, coal, coral, peat, oil and wood for two billion years and releasing oxygen to create the atmosphere that now suits us best. By burning too much fossil fuel and by converting limestone to lime for building and by clearing the forests that would go on 'fixing' the carbon dioxide we release, we are about to double the quantity of carbon dioxide in the atmosphere thus warming the globe and changing the climate.

dioxide humanity has pumped into the atmosphere since we began fuelling our industrial expansion with, first coal, and then oil. We should regard the sea around us with a new awe. Such an exchange rate cannot have been normal in the last hundred centuries before the greenhouse effect started to accelerate: if it had, then the seas would have been responsible for a steady, annual loss of carbon dioxide from the atmosphere. The temperature would have begun to drop, inexorably. The polar ice-caps would have moved south again. Such human civilization as existed would have been very different. Until the beginning of the industrial revolution, the ocean-atmosphere carbon exchange must have been more or less in balance. The deep, unheeding oceans seem, somehow, to be responding to a crisis we have engineered for ourselves, and reacting to soften the blow the future holds for humanity and the terrestrial world. It is as if the seas were taking up at least part of the extra carbon so as to cushion the shock to come.

But that, of course, is not what is happening. The ocean is a gigantic chemical retort, in which thousands of complex and sometimes unexpected chemical reactions are taking place on a scale which defies comprehension. The impact of sunlight on sea water turns some of it into hydrogen peroxide: the bleaching agent which turns hair 'Marilyn Monroe' blonde. The phytoplankton, the tiniest green plants in the sea, for instance, excrete about half of the world's sulphur as dimethyl sulphide, and that, too, might play its part in influencing the greenhouse world, not necessarily for the worse. (It could play a part in the formation of raindrops in clouds; thus changing the reflectivity of the clouds, damping the heating effect.) The ocean is also a source of methane: another greenhouse gas which is steadily increasing in the atmosphere because of human activities. Although the quantities of methane are small, each methane molecule is 20 or 30 times more effective at trapping heat than a molecule of carbon dioxide. There may be multiple troubles brewing in the brine.

But in the chemistry that concerns us here, in the first few metres of the deep, and at that infinitely thin layer which washes the air, and which in turn is caressed, soothed, whipped and driven by it, the ocean is simply reacting to the extra carbon in the atmosphere. The more carbon dioxide in the air, the more the sea can take up. But this reaction is only effective up to a point: and that point could be reached quite soon. The process is limited by how well the sea 'mixes': that is, how swiftly the CO_2 saturated water can be carried away so that the process can go on. Cold seas not only absorb carbon dioxide better: they are denser than both the ice-shelves along which they wash, and the warmer waters of the temperate and tropic zones. So they sink. Sink may not be the right word. The cold seas of the north turn into vast submarine waterfalls that dwarf Niagara and make the Amazon seem a trickle. Off the Denmark Strait, between Iceland and Greenland, the northern seas tumble downwards, taking 5 million cubic metres of water a second — 25 times the Amazon — down a fall of 3.5 kilometres. Then it carries its burden of carbon well below the level at which the day-to-day biosphere ticks over, and flows, in giant silent cataracts, to upwell very gradually somewhere in the tropics, to be warmed by the equatorial sun, and to then release some of the carbon dioxide.

Meanwhile, the cycle goes on: the warmer waters that have replaced them at the poles begin to cool and absorb carbon dioxide. That is the first point to remember. The reaction is dependent on temperature. It works best in a cool sea. The oceans are the world's great heat trap: the top 5 metres of the seas hold more heat than the entire atmosphere. World-wide, the oceans are warming at the rate of about 1°C a decade. That means that every year they become fractionally less efficient at absorbing carbon dioxide. What happens when the greenhouse effect really begins?

If the carbon dioxide levels in the atmosphere double in the next 50 years — and it will if the economic world of humankind airily goes on behaving as it has — the polar regions could warm up by up to 12°C. Most of that warming will be in the winter rather than the summer. The worst possible thing could then begin to happen. The oceans could surrender their role as a net absorber of carbon dioxide. Instead they could start releasing it. That is, instead of soaking up half of our carbon dioxide waste, they could start adding, say, 3 billion tons of their own to the atmospheric pool. This of course would start to make the world warmer still. A warmer world would mean warmer oceans which could mean that even more greenhouse gas was pumped into the atmosphere from the seas. Since 95 per cent of the biosphere's carbon is banked in the sea, the process could just go on and on. On the very simplest model, there is no reason why it should stop.

There are other elements which drive the spiral towards an overheating world. There is the albedo effect. Albedo is a measure of how much light is reflected. Something white — high albedo — reflects solar radiation, something black absorbs it. The white tops of clouds reflect sunlight: they cool the earth below them. Tropical rainforests absorb the sun's energy rather well; they are a heat trap. Arctic ice reflects it brilliantly. The poles are a cold store. That is why, on a baking day in the hot Alpine summer, the rambler is struck by pockets of snow and ice, usually somewhat shaded, that have stubbornly refused to melt. To a certain extent, it is the albedo effect that keeps the poles capped by ice. Oceans, on the other hand, being blue, absorb heat quite well. Just as the ice fends off the sun, the sea welcomes it, and quite literally warms to it, gathering its benison at one point and releasing it at another, which is why Britain, sluiced by the Gulf Stream, has a milder climate than its latitude dictates. But in a warmer world, the ice-caps would inevitably start to shrink. That would mean more dark areas to absorb the sun's heat and give the whole process yet another kick forward. That is what is meant by positive feedback.

But there are other feedbacks, each adding further momentum to the (so far) mercifully hypothetical headlong rush towards an incinerated world. Arctic and Antarctic waters are heavy and urgent with life. In fact, the polar seas are far richer in living creatures than most tropical waters. Life there, as in warm climates on land, depends ultimately on green plants that develop by using photosynthesis. This is odd, because for six months of the year, there is hardly any light at the poles, and for part of that time, and for part of the summer, the light comes in at such a low angle that it is hard to imagine that it can be much use to a plant. The fact that the plants — the phytoplankton, the beginning of the marine food chain and the basement and foundation of life — are in the sea, and the sea is thoroughly covered with reflective ice, and that in the polar spring, when (to all intents and purposes) the weather is as cold and as unstable as winter, makes it even more difficult to understand.

And yet, precisely because of the ice, there is a little, hidden miracle every Arctic winter. This is powered by yet another strange property of frozen water. It can propagate light downwards. Imagine the sub-polar semi-darkness; imagine the sun so low in the sky that its rays simply graze the sea ice, touching it with a glancing blow before skimming away into space again like a flat pebble on smooth water. But not all of it glances away. Some of it is trapped and channelled by the ice, straight downward, there to illuminate the cavernous ceiling of the closed world at the bottom of the floe: the world where, thanks to water's unique property of freezing from the top down, life goes on, and, thanks to the propagated light, so does photosynthesis. The light is weak, but it is enough. Algae colonize the base of the ice: they play the role that, in open seas, would be played by the algae that grow on sea-bottom sediments. They appear on the underside of the ice as it forms in the autumn, and they provide a source of food through the winter for the zooplankton, the smallest carnivores in the sea, and bigger predators and the Arctic cod and through them the seals, walruses and polar bears and the peoples who settled the Arctic Circle. The tiny algae are as it were the blue touch paper, sizzling through the long polar night, keeping life in business until the explosion of surface phytoplankton and other marine organisms when the sun returns in the spring.

Thus it is the ice itself that keeps the Arctic alive: without it, the biological spring would still occur: but it would happen much later in the short polar season. The consequences for the Arctic ecosystem would be considerable. But so would the consequences for the whole globe. For the green plants of the sea 'fix' carbon dioxide dissolved in the sea from the atmosphere as surely as, and for even longer periods than, do trees on land. Some of it, of course, is returned to the biosphere quite swiftly: such microscopic creatures do not live for long. But much of it goes into the food chain, and a substantial proportion of that ends where it began, in the sea, drifting slowly downwards to the sea-bottom in the form of decomposing organisms or tiny skeletal structures, to be stored for centuries or millennia or aeons, to be returned perhaps, one day when the continental shelf has tilted, or the sea has retreated in another ice age, and be exposed to the air again. In just this way the limestone uplands of Yorkshire and

Pennsylvania rejoined the biosphere, some of them still studded with corals and crinoids that betray an origin in ancient tropic waters, slowly surrendering once more the carbon in their calcium carbonate as they erode under the wind and rain, or are quarried for building work, or steelmaking, or for absorbing the sulphur dioxides from coal-burning electrical power generating stations, and undergoing conversion to gypsum.

Nobody knows quite how much carbon the phytoplankton account for: their capacities for growing by fixing carbon dioxide are not limited by how much dissolved gas is available in the sea, but by the other nutrients they need, such as phosphorus. They bloom — that is, flower so thickly they discolour the sea — in areas where there is an upwelling of nutrients from the deep, or a run off from the rivers on land. The richest, most productive waters of all are off the Antarctic continent, but there are bursts of abundance off the Chilean coast and the Gulf Stream. There have been attempts to measure how much carbon dioxide the phytoplankton take up as they grow. If the conditions are right, the sea's almost invisible green fields can fix a kilogram of carbon a year over an area of 3 square metres. There have been attempts to quantify the sea's carbon. The surface waters — which are continually exchanging carbon with the air — might hold 500 billion tons. That is rather more than is temporarily trapped in vegetation on land. But the deep ocean, which can hold its burden of dissolved carbon dioxide for tens, hundreds or perhaps thousands of years, or its organic or skeletal carbon for what, for our purposes, is forever, could be playing host to 35,000 billion tons.

Any process that triggered the oceans into beginning to surrender more carbon to the atmosphere than it took from it in the first place would be calamitous: calamitous, because the seas have so much more to surrender. And the attendant warming would feed the flame that could go on burning until, like the planet Venus, all signs of life would be obliterated: and the earth too, would have a surface upon which lead would melt, upon which water would instantly flash to vapour. Alongside that prospect, the spectre of the Houses of Parliament colonized by kelp, sea urchins and nudibranchs, her corridors patrolled by conger eels, sounds like light relief. But either conjecture contains within it a multitude of ifs and buts. All scientists have been able to do is play with models: only within the last two or three years has there been a serious international attempt to begin to answer some of the larger, simpler, starker questions about ocean-atmosphere processes.

But each large question takes ships, satellites, sensing and sampling instruments, computing power, software, time, energy, intellect and money to frame, let alone answer: were each to be answered, a hundred smaller questions would immediately pose themselves. Oceanography is a relatively new science: the sea is old, deep, wide, vast and contains almost everything within it. To interrogate the sea is to ask it questions that are as old as time and as all-embracing as Creation. And each question must be asked of the four winds as well. 'Hath the rain a father?' asks the Lord in the Book of Job. 'Or who hath begotten the drops of dew? Out of whose womb came the ice? and the hoary frost of heaven, who has engendered it?'

During 1988, when it became clear to the world that greenhouse warming was inevitable, and the problems inherent in it apparently intractable, there was an almost universal consensus among European, American and Soviet scientists that the key to most questions lay in the oceans. Britain responded by sacking a disconcerting proportion of its government ocean and marine scientists, and breaking up the team that had identified one of the most startling questions of 1987: why, over 25 years, has the average height of the waves of the North Atlantic risen by 25 per cent, when the average speed of the winds remained constant. It is as if the bleak waters of our own shores, implacable but also nurturing, have already begun to rise in a murmur of anger.

Meanwhile, the world has other questions to ask of the waves. Only in the last decade has science begun to identify the scale of the interdependence of winds and ocean, climate and weather. For even the weather isn't just something that blows in the wind: that too, ultimately depends on the oceans. Northern Europe is warmed by the very waters that seem so hostile: 30 per cent of the heat the North Atlantic receives from the sun is released there. Slight shifts in the warmth of the Indian Ocean determine whether people starve in the eastern Sahel; if the tropical Atlantic is a little warmer, then it rains in the arid zones south of Equatorial Africa. These shifts themselves are a function of the unimaginably vast submarine currents that are as veins and arteries in the body of the ocean: and the currents are as they are because of the temperature differences that exist between pole and equator. If those differences diminish — and if the globe warms unevenly under the greenhouse effect they must diminish — then the alarm bells should start ringing everywhere. The consequence can only be a series of large scale sudden shifts in climate. They will be sudden because the evidence of the past — only in the past few years has there been some kind of confirmation from the Greenland ice cores of a link between carbon dioxide and climate change before and after bygone ice ages — offers no other conclusion. The last time it happened, 10,000 years or so ago, the globe may have gone from cold to mild in 20 years.

It is as if climate shifts in the way rock does in an earthquake zone. By satellite, you can measure the relative positions of Phoenix, Arizona, and Los Angeles. The former is moving south a little all the time. The latter is shifting perceptibly north. But grains of soil, the houses and the telephone poles on either side of the geological border, the San Andreas Fault, stay in the same position, relative to each other. It is as if the rocks of the two tectonic plates that bear each city are deforming elastically under the strain. One day they will snap, catastrophically, into the positions the rest of the Eastern and Western America dictate. One day too, the invisible inertial forces holding the climate zones of the world in what we think of as the proper place will also snap. The climates will begin to shift. Unfortunately, there is no evidence that they will stop shifting. That is because there are also feedback forces at work on the land. One of them, of course, is us.

CHAPTER 4

On the beech

In 1987 a Cambridge botanist began exploring the herbarium in his own university. He examined the leaves of oaks, beeches, maples, poplars and hornbeams collected and dried since 1750 AD (conveniently *before* the start of the Industrial Revolution), and put them under the microscope. He found that the leaves collected in this decade had 40 per cent fewer stomata in each square inch of surface than those collected 200 years ago. Stomata are the pores through which plants breathe and release water.

The trees had, in effect, detected the greenhouse effect and responded to it. As levels of carbon dioxide started to go up by 20 per cent the density of stomata started to go down by 40 per cent. The trees became more 'efficient'. Just as the ocean responded to the extra pressure of carbon dioxide by taking up more than it released, the terrestrial plants too began taking advantage of the extra investment currency being circulated by the great global Friendly Society. They spent less energy acquiring carbon dioxide, and used water more efficiently. They had the opportunity to make more foliage, more flowers, more fruit; to colonize the land more lustily. Crops — or at least some of them — will welcome the extra greenhouse gas; it is estimated that, with a doubling of carbon dioxide, cotton will increase its yield by about 100 per cent; sorghum by more than 70 per cent; wheat and barley by more than 35 per cent. Soya beans, rice and maize will all do better. It is an example of negative feedback; an instance of the biosphere's capacity to regulate itself. Even if you don't believe in the ideas of a benign Nature, or a caring deity, it still looks wonderfully like part of a divine plan, or at least an unusually subtly-designed mechanism.

That is the good news. It also suggests an obvious way of artificially damping the greenhouse effect, by planting more trees to soak up the three or four billion tons of carbon we dump into the air each year. It means doing some global-scale calculations. Numbers are tricky things: the bigger they are, the less they mean. Altogether the plants of the world — in the oceans and on the land — use photosynthesis to fix roughly (very roughly) 100,000,000,000 tons of carbon a year from carbon dioxide. Appalled at the idea of presenting the public or the politicians with a figure with 11 zeros attached to it, one international team of scientists involved in the calculations tried to convert it to something which could at least be imagined. They came up with the image of a coal train, filled with carbon which went from the earth to the moon and back, five times. Two thirds of the train would be filled with carbon from the phytoplankton of the oceans. About a third would bear the photosynthetic burden of the trees, mosses, liverworts, shrubs, weeds, grains, vegetables and grasses. While there is not a lot we can do about phytoplankton in the oceans — we don't know enough to even begin to think of managing them — we can certainly manage the vegetation on land. It's a skill as old as Adam and Eve.

Gregg Marland of Oak Ridge National Laboratory in Tennessee did the sums. He calculated that if you planted trees which would grow at the speed of the American sycamore in the south-eastern United States they would take up carbon at the rate of 7.5 tons per hectare every year. An area of about 700 million hectares would fix 5 billion tons of carbon every year. In the USA alone, he further calculated, there were 70 million acres of unproductive land with the right rainfall for such plantations. They would, of course have to be managed, harvested and replaced, and the wood used for building or furniture. Plantation on this scale, he argued, would buy time for the world to overcome its addiction to fossil fuels and think of some other way of tapping energy.

There are, of course, other reasons for planting trees by the thousand anyway. They protect water catchments and slow down water run-off; they shelter wildlife; they provide windbreaks and halt soil erosion. They have a role in heating and cooking for the poor in the hungry nations; their fruit and nuts provide food for humans and their leaves fodder for cattle; they provide housing, furniture, weaving materials, and paper. Some of them fix nitrogen directly from the air and enrich the surrounding soil. Choose the tree and you can also harvest silk, honey, wax, turpentine, resin, essential oils, rubber and even natural medicines. It's worth saying again: trees provide food, shelter and fuel; they conserve the soil and the water and they could save civilization from heat death all at the same time. If you plant enough of them in the same place, they can even create their own microclimate, and keep themselves secure against drought by recycling their own water; that is to a certain extent what happens in the Amazon rainforests. It sounds like nature's ultimate bargain basement offer.

In fact, as everybody must now know, an enormous area is being torched, chopped, slashed and bulldozed within the tropical forests every year. Ninety per cent of Atlantic forests of Brazil have gone; so much of Indonesia and the Philippines has been cleared that there is drought and soil erosion in what is still technically, monsoon country. The vast forests of West Africa have shrunk, and are ringed by scrub and savannah, slowly desiccating as the people stand helplessly by. This is being done despite ostentatious wringing of hands and condemnation by the environmentally concerned in Western Europe, which, of course, disposed of almost all its forests hundreds of years ago, and is now subjecting its remaining woodland to acid rain, lead pollution and photochemical smog.

The clearing of the forests is of itself accelerating the greenhouse effect. An astonishing 90 per cent of the carbon in terrestrial vegetation is locked up in wood. The clearing of the jungle is releasing anything from a billion to 2.6 billion tons of carbon dioxide into the atmosphere: some of that is let slip not as the trees char or rot, but when the soil that once held them is ploughed. The tithe of the Amazon is fed into the heat machine every day. But the heat machine may not be self-correcting, it may be another example of dangerous positive feedback. It works like this. All trees are to some extent sensitive to temperature. Left to their own and Nature's devices, they tend to flourish in a terrain and climate that suits them perfectly, and wilt and go under if the same terrain suits competitors better. Since the range of temperatures in the tropics is not likely to vary greatly under global warming, that is where the extra carbon dioxide is likely to do most good. That is, however, exactly where the forests are disappearing fastest.

But in the temperate zones, the species that respond to a winter chill and a summer swelter are going to be under stress: they are going to have to pick up their roots and trudge north. It has happened before: trees have marched south and north before and after the Ice Ages. Fossil pollen records provide a fairly clear picture of the ranges of trees almost everywhere. Under the ice regime that ended 10,000 years ago, the forests migrated at the rate of about 20 kilometres a century. That may not seem very fast, but even those that produced light seeds designed to be dispersed by the winds are not at a great advantage. Studies have shown that 95 per cent of all grass seed falls within 9 metres of the parent plant. Fewer than 5 per cent of the seeds of the Engelmann spruce have travelled more than 200 metres down wind. But of course, the forests cannot migrate. Industry, mechanized farmland, asphalt motorways and waste dumps — block the path almost everywhere in the temperate world. Nor could trees migrate fast enough. A warming of 3°C would require the forests to shift 250 to 350 kilometres north, or 500 metres up a mountain. In a mixed forest, some species would survive: others would go under. The character of the glorious woods of New England is already under threat. There is another factor. Warming doesn't affect the rate of photosynthesis but it does affect respiration, in which plants take up oxygen and release carbon dioxide. One scientist has calculated that if the overall global carbon in circulation from all natural sources is 100 billion tons, then the warming so far this century could be delivering an additional 1 to 6 billion tons of carbon a year to the atmosphere in increased respiration. And the warming so far has been about half a degree.

An increase of 2°C or 3°C is a lot of heat. The last time it happened, according to the fossil record, pawpaws bore fruit where Toronto now stands, and tapirs wallowed in North Carolina. But a dramatic vegetation change does not require a 3°C warming. About 180 years ago in the eastern United States the dominant trees were conifers: the spruce has now been all but extinguished by the beech. This has happened in a lifetime: the lifetime, that is, of a tree. Prehistoric Europe has paid a price in species variety under changing climate regimes. Once in France and Germany there bloomed the watershield, the sweet gum, the tulip tree, the moonseed, the hemlock, the arbor vitae, the white cedar. They, and others, ringed the North Pole. Then came the Ice Age. In North America, the trees were able to move south before the slowly advancing glaciers. In Europe, their retreat was blocked by the Alps, the Pyrenees and the Mediterranean and they perished, leaving only their pollen in the peat bogs. As the average world temperatures begin to soar, victims will collapse on the march north. There is a dwarf birch which can only grow where the temperature never exceeds 22°C: we may see it go.

The temperate forests that survive will have to confront two stresses. Either the climate will become drier, in which case some species will fail or it will become wetter, in which case other species will dwindle under the change. Those that don't will still be under assault. A drier climate means a greater risk of forest fires, each blaze spilling more carbon dioxide into the system, and stoking up the heat. A moister climate won't necessarily please the survivors, but it will certainly encourage the insect pests that prey upon them. The warmer and wetter it is, the quicker they develop, the faster the spread, the bigger their range. If the wind patterns change as well — and they are almost certain to do so — the locust and the army worm could have new fields to conquer. The screw worm cannot survive where the temperature drops below 10°C. Global warming opens more of the world to its destructive jaws. Either way, the temperate forests will be under pressure. So will what are called the boreal forests — the sparse, ghostly needle-leaved trees and birches of Siberia, Northern Canada and Scandinavia. The warming of the tundra will open new horizons for them, and their geographical range will increase. But all the pressures that afflict the temperate forests will land on the forests of the far north with great force. One projection — and it is only a projection — suggests that a 3°C rise in temperature could mean a 37 per cent decline in forest area.

This is because, of course, the rise is uneven. Take the case of temperatures in Niwot Ridge, Colorado in 1970 and 1971. The warmer year was 1970. The average for the whole year was an increase of 1.7°C higher. But the absolute minimum temperature — that is, the coldest it became — that year was a whole 7°C higher. And then remember that in the high latitudes the average warming could be 10 or 12°C, and that the absolute minimum in a number of regions could be even greater. Apply that heat to fragile plant communities of the tundra, and watch the more opportunist species of the north forests march in and take over. This is not hyperbole. If the greenhouse effect is really happening, we will be able to observe it within 10 years in the tundra. The frozen bogs will retreat by at least 4 degrees of latitude, the boreal forests will invade by 10 degrees. The ecological cost in the warming of the north could be colossal; caribou and muskox and polar bear all need the security of the classic sub-Arctic winter. They need the firm footing of the ice bridges to carry them to new food supplies; and the caribou and the ox can't forage in wet snow which has then frozen. When disaster overtakes the caribou population, it does so in warm winter, not a cold one.

But the collapse of the permafrost imposes another positive feedback: about 14 per cent of the world's soil carbon is stored in the peat and the frozen bogs, which melt briefly in the polar summer for a few inches and then freeze again. The carbon has been accumulating as litter in the subarctic for 10,000 years. As the ice cover melts and trees and annual plants begin to colonize and disturb the precarious stability of the region's soil, that, too will rejoin the great carbon currency exchange, and reinforce the big heat, and see to it that the warming of the Arctic goes on.

But that is not the end of the feedback. The agony is piled on. Even a relatively small rise in sea levels will actually reduce the area of land available for plants and animals. There will be several effects. One of them is erosion. The rule of thumb is that in land protected by shallow-sloping sand dunes, a metre rise in sea level will sweep away 50 to 60 metres of coast. Where there are barrier islands, they will be drowned. One of the likely consequences of warmer waters and changing weather patterns is a noticeable increase in the frequency of storms. Under present regimes, storms can have calamitous impacts. The salt water that invades and alters the geographies of the great estuaries will do so with greater frequency, and increased bitterness. The impact will be even worse on those deltas fed by rivers which have been dammed. These are sinking because the silt that would have renewed them has been trapped upstream. The mighty Mississippi delta is sinking at the rate of 2 millimetres a year. Venice, the Po delta, and the Gulf of Cadiz are all at risk.

Some deltas are still growing. The mud still pours into the mouths of the Irrawaddy in Burma, the Ganges and Brahmaputra in Bangladesh, and the Mekong. Without our interference, these deltas would go on growing: the silt is naturally held and detained by mangrove swamps. But the mangroves are being cleared, by fish farmers, by land reclaimers, by the assault of toxic pollution. Coastal land everywhere is becoming vulnerable: Indonesia — an archipelago with a series of shorelines so serrated they could be used as illustrations for a textbook of fractal geometry — has 15 per cent of the world's coastline. This means, of course that a greater ratio of its land is close to the sea. In fact at least 40 per cent of land surface is now considered vulnerable to rising sea levels.

The argument goes that a 1 metre sea level rise could affect all land up to the 5 metre contour. A 1 metre rise is higher than the more careful scientists predict, at least by 2020, but it isn't unreasonable. In the last 10,000 years since the Ice Age the seas have risen by 65 metres. From that perspective, a metre doesn't sound much. But within the 5 metre contour lie the homes of a billion people — and one third of all the cropland in the world.

Clearly, even if the world's population was stable there would, as a consequence of rising sea levels, be greater pressure on inland soils: the land which new forests might be permitted to colonize would be needed to establish new agriculture and new cities. But the area available either for trees or people is shrinking: 15 million acres every year turns to desert. Another 50 million acres every year become too poor, too arid, too sterile to support crops or cattle — because of aggressive farming, over-grazing or salinity because of bad irrigation. Another 800 million acres is losing topsoil at an alarming rate because of erosion either by wind or water. Worse, the impact of global warming, while it will certainly bring rain to some areas, is likely to hit hardest at some of the world's most productive land. Until now, the plains of America have been almost perfect for growing corn. For decades they have underpinned the security of the world's food supply. But, periodically, they are hit by drought. It happened in the 1930s and the 50s, and notoriously in 1988, the year the greenhouse effect was declared officially as an item on the geopolitical agenda. The plains are, ecologically speaking, on

. the edge. They too could tip over as the world warms, becoming a home for neither buffalo, nor corn, nor people, nor trees, gradually surrendering their soil carbon to the desert air and the fires of global warming.

Thus even if the world's population stabilized right now, the inland and higher soils would be under pressure. There would be dangers of more erosion, more aridity, more overgrazing. But the world is growing fast. There are another 90 million mouths to feed every year. At some time in the next century the world could number 10 billion souls. And each new soul contributes to the positive feedback in other ways. As the developing nations struggle to achieve a modicum of prosperity, or even economic security, they too start releasing growing amounts of carbon through the use of fossil fuels. They, too spread the nitrogen fertilizers that may be ending up in the skies of nitrous oxide at the rate of 12 to 15 million tons a year — nitrous oxide is 200 times more efficient than carbon dioxide as a greenhouse molecule. They too invest in chlorofluorocarbon (CFC) plants to build refrigeration systems to preserve the food they so badly need, but every CFC molecule has 20,000 times the global warming impact of a molecule of carbon dioxide. And at the end of the chain, the cost of decomposition. From the rice paddies of those nations only a few steps ahead of famine, the natural gas methane wafts to the sky at the rate of up to 170 million tons a year. It joins the 100 million tons belched and farted every year from the intestines of sheep and cattle; it joins the 70 million tons oozing from landfill sites from the rich world's wastes; it joins the 40 million tons wasted by the oil and gas industry and the 55 million tons from rotting vegetation.

Methane is 30 times more effective at heating the world than carbon dioxide. Even worse, the levels of methane in the atmosphere are rising at more than 1 per cent a year: this gas should, to a certain extent, be taken up by soils. This is not happening. It isn't happening, the latest experiments suggest, because the increased use of nitrogen fertilizers is blocking the process. So the world is caught in another double bind, another case of feedback. Both methane and nitrogen fertilizers add to the greenhouse effect, and the use of nitrogen makes the methane problem worse. Methane is beginning to worry the climate scientists a great deal. Enormous quantities of it are stored in the frozen, compressed soils of the tundra and the northern forests. It is stored in a hydrated form, that is combined, under pressure, with water. One calculation has it that a cubic metre of water in the near Arctic might contain 170 cubic metres of methane. Once the tundra starts to warm, it will release not only its carbon dioxide but its methane as well. The belt of soils round the Arctic contains colossal quantities of methane, and it is precisely there where the warming will be at its worst. Once the gases begin to seep into the atmosphere, the cycles of feedback will accelerate.

The CFCs will go. They are destroying the ozone layer, and could be completely banned from use under a United Nations protocol by the year 2000. But methane will go on rising. It may end up being responsible for 20 per cent of global warming, but how do you legislate for bacteria in a rice field? Or a cow's digestive tract? How do you stop a termite farting? Or smother an aquatic compost heap? And how, when there are 86 million new mouths to feed every year, do you control the nitrogen fertilizers which make up 6 per cent of the greenhouse effect? The auguries for the world's forests are not good. The auguries are not good for anybody, except the insects.

CHAPTER 5

Fuel's gold

The hummingbird spends 75 per cent of its time sitting on a perch. When it does fly it flies so fast you can't see its wings move: its method of feeding — hovering in front of a blossom while it sips nectar with its long bill — is one of the most hectic in creation. Small creatures are at a disadvantage in the energy handicap stakes. The smaller the animal the faster it sheds heat through its skin, because the body mass is small and the surface area large in proportion to it. Because of this, and because of its exhausting feeding strategy, the hummingbird, ounce for ounce, burns energy at roughly 30 times the rate of a human. If it dined off cornflakes or cold potatoes, it couldn't survive. The hummingbird needs energy in its swiftest, highest octane food form: sugar. It needs it fast and often and it needs to digest it quickly and waste almost nothing. Even though it spends three quarters of its waking hours sitting on a twig, it still needs to imbibe 180 meals a day: three times its own bodyweight in nectar syrup. The Anna's hummingbird must be the ultimate sugar digestion machine: it can extract 97 per cent of the energy from glucose and it will have begun to dispose of the waste in its faeces within 15 minutes. When it is not actually eating, it has to sit still: it has to divert energy for digestion and it has to save some of its strength to get it back into the shrubbery to fill its crop again. It has, to fall into anthropomorphism and teleology and all the things biologists must avoid, selected from the entire range of energy efficiency strategies available to creation, and opted for a short life and a sweet one.

There are other examples. The elephant, for instance, lives long and takes it slowly: consuming enormous quantities of low energy cellulose in the form of grasses and leaves, and it too has an energy problem but of a quite different kind. Because of its vast mass and correspondingly smaller surface area, and because it lives in latitudes where the average temperature is quite high, it needs to lose heat rather than save it. That is why it has big ears, giving it an extra few square metres of skin through which the body heat can flow to the surrounding air. Noticeably, the African elephant, which lives in an (on average) warmer climate than the Indian elephant, has the smaller body and the bigger ears. It is both possible and useful to think of all living things as energy-strategy solutions. Siberian tigers are bigger than Bengal tigers because they need to conserve more energy to keep them alive and warm between kills every two or three days. If the only food around is the aphid then the coal tit has to eat one every 2.5 seconds right through the daylight hours or it starts to lose the race to stay alive. It is small, and doesn't have much fat in the bank.

The whale, creation's giant, doesn't just have fat in the bank: it keeps it as a thick layer of insulating blubber to slow down heat loss in the water. In the Antarctic or the Tropics, water is a good conductor of heat away from a body: about 90 times more efficient than air at the same temperature. Without the blubber and its initial bulk, the slow feeding whale wouldn't be able to survive. Thus the whale has made its own deal with the energy bank, and the smallest size at which what scientists call an aquatic endotherm can be born has been calculated at 6.8 kilograms. This is in fact about the birth weight of the smallest whale, the river dolphin. Other creatures have energy loan strategies: the polar bear can spend a few autumnal weeks gorging itself at the rate of two million calories a day and then shut down, slow its metabolic rate, and hibernate for up to seven months, neither eating, drinking, urinating or defecating, its heart pumping at a mere seven or eight beats a minute instead 50 or more while its somnolent body lives off the huge reserves of fat. The arctic warbler; just before it migrates can double its weight for its travelling expenses and lose it again on the journey. Polar bears and arctic warblers and whales live on the margin: they survive by energy efficiency.

So do humans on the margin. Puzzled by the energy budget of some African women — how can they walk for miles in the heat, bearing huge loads of firewood or water, when they are metaphorically speaking, but steps away from malnutrition or starvation? — scientists persuaded some Kikuyu and Luo women to pace a motorized treadmill, carrying a weight handicap. They measured their oxygen consumption rate because the speed at which you consume oxygen is a measure of the energy expended. They found that the women could carry 20 per cent of their own bodyweight without increasing oxygen intake. When they carried 70 per cent of their own weight, the oxygen budget increased by about 50 per cent. Trained soldiers carrying backpacks don't do as well. A 20 per cent load increases oxygen demand by 13 per cent; a 70 per cent load makes them breathe almost twice as hard. Clearly if the atmosphere is a bank, a stock exchange, a commodities market, then the ultimate common unit of currency is not the crumpled, grubby pieces of paper or soiled coin we call money, but energy. It arrives, in its raw form, free from the sun, but to use it — to be born, eat, sleep, breathe, digest, make love, rear children, write books or climb mountains — we have to work for it. We buy it and we spend it, and the more we spend the more we pay for it. If animals have a notion of economy, they must think in units of energy. So, at least unconsciously, must humans living on the margin.

That is why the greenhouse effect is a cause for alarm. It means an enormous shift in collective economic and political thinking. It must be thought about in the currency of energy units. Some animals build houses, use tools and store wealth in the form of food but even for those, the trading arrangements with the biosphere are relatively simple. They take what they want when they need it: they grow, play, eat, sleep in the sun, educate their cubs or nestlings, run from enemies and then die. Only humans build roads, and then build motor cars to drive on them, count their wealth in personal stereos, electric toasters and executive toys, and measure their security in intercontinental ballistic missiles and satellite warning systems and use economic growth as an index of contentment. The wealth may seem to come cheaply. Just as someone who accidentally finds a sizeable lump of gold is instantly rich just from the effort of stooping and picking it up; so a nation that strikes oil is transformed from a poor, arid, grubbing kingdom into a major player in the geopolitical poker game, but in both cases creation has invested energy, and will exact its price. The gold may mean a year or two flying around the fleshpots aboard Concorde; and the oil may mean the transformation of a people from a few huddled black tents into a skyscraper 'paradise', but in both cases the financial wealth only has meaning if it can buy somebody else's effort, and this effort is ultimately expended in carbon dioxide. The cost comes back in global warming which threatens the security of the rich.

That is why ecologists and economists and climatologists are worried. They have just begun to realize that the lion's share is just that — what a lion needs to survive. Economists and ecologists, like everybody else, want foreign holidays and personal stereos and comfortable, air-conditioned cars, but they now know that for hundreds of years we have been counting the costs of them in the wrong currency. Instead of pounds or dollars or yen, it should have been in units of energy expended. Every detail of the appurtenances of wealth — the rolled steel, the wires, the plastic flex, the design of the dashboard, the instrumentation, the steering wheel, and the silly sticker saying My Other One's A Porsche, or the fast forward and the playback mechanism, and the foam rubber padding on the headphones — had to be paid for in oxygen consumed, carbon dioxide exhaled. Fuel is just another form of gold: we should spend it carefully. Energy is just that: money to burn, with the heat being dissipated into the atmosphere.

Sometimes the sums are easy. Make some cement. To do it you convert limestone, otherwise calcium carbonate or $CaCO_3$ into lime, CaO. You can see what is missing: CO_2, carbon dioxide spiralling from the cement works straight into the atmosphere from whence it came when the limestone was laid down in the oceans millions upon millions of years ago. With every 7 tons of cement you also release a ton of carbon dioxide. When you build a skyscraper or a motorway, you build trouble. Burn some coal. Or oil. Or natural gas (methane). Or wood. They are all hydrogen and carbon based. In each case, a chemical reaction takes place, oxygen is consumed and converted with the carbon to carbon dioxide. We use the heat of the reaction and are left with some ash and some gases, one of which is carbon dioxide. Even nuclear power, which produces no carbon dioxide, because it rests not on chemical reaction but the fission of unstable atoms, rests on a bed of cement and burned hydrocarbons.

The energy to make the concrete and assemble the plant was supplied by fossil fuels, more of the same energy must be spent handling and disposing of the unwelcome radioactive wastes, and 20 or 30 years later, when the plant has to be decommissioned, other sources of power will be needed to spend up to 100 years dismantling it. There are no free lunches under the sun.

The crisis managers of the world's developed nations now know this very well. They know that there won't be a simple answer to the greenhouse effect because in the first place, it's almost certainly unstoppable. The most they can do is to stop it getting very much worse, very fast. They also know that, with an investment of effort, ingenuity and a switch of national resources, it should be possible to use the energy we produce now more efficiently: that is, to get more value from each unit of power we produce, and use that to finance our economic growth. This has happened in most of the richer western nations since the 1973 oil crisis, when the oil and petroleum exporting nations decided to unite and impose a stiffer price upon the stuff that makes the world go round. But this time it is different. Then we measured our use of oil in money. This time we have to start measuring the cost of everything in energy, and the value we set on the energy will have to be measured in the release from that energy of methane, carbon dioxide and nitrous oxides.

There are other options open to sophisticated, wealthy nations. They can switch from fossil fuels to less polluting energy sources. Coal is a relatively filthy way to power the world; oil is better; natural gas is better still, but they all yield carbon dioxide. Other sources are even cleaner. A nuclear power plant, once up and running, releases no pollutants at all, unless something goes wrong (in which case the potential for long term environmental damage from a deadly rain of caesium 137 or strontium 90 is enormous). Hydro-electric power, once you have discounted the energy cost of building the dams and making the turbines, taps the energy of river water on its headlong journey to the sea; tidal barrages can exact an electric charge from the twice daily surge of the sea; there are schemes for exploiting the natural energy of the waves; for plugging into the wind; for running smaller power stations off the difference between the warm water at the surface of the ocean and the cold below; and of course solar power technology focuses the energy of the sun directly. And, of course, there can be a greater investment in forestry. One American power utility has already shown a stylish lead by planting an area of forest big enough to soak up roughly the annual release of carbon dioxide during the station's lifetime. It isn't an entirely selfless gesture: the timber will have value; and the forest will enhance the attractiveness of the region; it will also be a shelter for wildlife, a catchment for water, a hedge against erosion.

But the steps that the rich nations take — if, indeed, they get around to taking them — will be as nothing. For decades, energy use has been growing at the rate of 1 per cent a year; lately, it has risen to 2 per cent. In 1989 at least one of the many groups of economists making forecasts decided that carbon dioxide levels in the atmosphere would double not by 2050 or 2030 but by about 2010 AD. The inevitable global warming could be here sooner than we had thought.

The problem is, of course, that nations are hooked on growth: only by economic growth have politicians believed they could fulfil promises of liberty, security and the pursuit of happiness. But the poor nations are growing even faster. Their economies may be in ruins but their populations are swelling. And in the race against a parallel swelling poverty, they too are rapidly industrializing. To keep abreast, the farmers of the poorer nations need trucks, tractors, fertilizers and irrigation. The fast-aggregating cities of the Third World need reservoirs, power stations, roads, transport, airports, television stations, refrigeration plants, hospitals and cars, and above all, jobs. All these things are now the difference between hope and despair. Moreover, the security and happiness for which the billions of poorer nations have been educated to hope is measured in the same things that provide indices of wealth and contentment in a rich world. The Mexican slum-dweller and the Zambian villager and the Chinese peasant farmer may be in want of a roof, a job, a doctor nearby, seeds, fertilizers and a sure supply of water, but they, too, dream of owning a car, a personal stereo, and a chance to take the children to Disneyland. Television, no less than foreign investment, and the march of the multinationals, has seen to that. A billion Chinese have had the chance to see *Superman,* and watch the British television show *Yes, Minister.* Zambians have seen Agatha Christie's genteel elderly detective Miss Marple catch the criminal in the plush villages of the English countryside and noted the paraphernalia of security: armchairs, the tea services, the houses with a bedroom for every person and the delicate, thinly-sliced cucumber sandwiches which nobody eats.

The difference is that there are more of the poor, and more to come. Within the next century, 95 per cent of all the projected population growth on the planet will take place in the developing world. The world's population will have climbed from 5 billion to 10 billion. The number of people who will need jobs in the developing world is, by 2025, going to be greater than the number of people who are in work everywhere on the planet today. The equation is quite simple: more mouths means more jobs, means more industry, means more energy use. Where they can, they will use their own fuel sources: those that have oil or coal will burn it. Those that haven't will buy it from those who have enough to spare, and if they have no money they will buy it with foreign loans that will keep them poor. Fossil fuel consumption will just go on rising. One calculation is that it could rise to 18 billion tons a year at some point in the 22nd century. If one third of all that stayed in the atmosphere, the carbon dioxide proportion could rise to 1500 parts per million. Just remember that it was about 275 parts per million before the Industrial Revolution; it is now about 320 ppm, and the politicians of the richer nations are worried right now.

The irony is that, on the whole, the wealthy nations have made their wealth at the cost of the poorer nations: they must now explain to the poor that they must not have the same aspirations, or it will be the ruin of us all. This may not impress the near-desperate citizen of an African shanty-town, whose life is a dreary trudge for water, maize-meal and firewood, and whose spare moments are spent looking for work, any work at all, and whose only escape is cheap beer and an evening at the cinema watching the suave fictions of Hollywood smash up expensive cars as if they were objects of no value, spend vast sums of money they seem not to have worked for, and order meals they don't seem to eat.

But in any case, the developing world has to run its course of economic growth. There is no choice. The developing nations have more or less stable populations because they are developed. Secure families in rich countries — families which are sure that their children will survive into adulthood, will get a good education and a good job, and that the parents will enjoy a stable and happy retirement on a generous pension — tend to have only a couple of children. Insecure parents in poor countries who fear that their babies will die early, and that those that survive will have to emigrate in search of work when they grow up, need large families to spread the risk and share the burden in the struggle for survival. That is why there is a 'wait and see' school of thought in the great greenhouse debate. We need, they argue, to find out what the greenhouse effect means, in detail, region by region. We need to start developing alternative energy sources. The rich nations need to practise energy efficiency and export the technology to the poorer ones. But the big decisions will have to wait until the developing nations have begun to catch up, and their birthrates have slowed. By that time, the argument goes, the world will have a clearer idea of what to expect.

The trouble is that, strictly speaking, the world won't have a clue what to expect by, say, 2050. A search through all recent climate history has only once recorded a world hotter than now. At a point between the last two ice ages, the average annual temperature of the globe was 1°C warmer than it is now. A simple doubling of carbon dioxide in the atmosphere will take us to exactly that point: one degree warmer. The rest of the warming to come is forced by the feedback effects already described. With a presumed 30-year-lag in the effect of carbon dioxide upon global warming, by the time we *get* to a temperature rise of say, 2°C, we will certainly be *heading* towards a rise of 4°C. No one can possibly know for certain what that will mean for agriculture, weather patterns, ocean circulations, desert spreading, forest growth or rainfall. We will be flying blind into a world we don't know anything about. There are other factors. Carbon dioxide is not the only pollutant, and the greenhouse effect not the only future to fear.

But the big catch with the 'wait and see' strategy is that the developing world is in economic fiefdom to the richer world. If the poor are to catch up, the rich nations, collectively and willingly, will have to surrender their economic advantage and forego the luxuries of plenty so that poor, unknown peoples very far away can join the banqueting table at which plain sensible fare is spread evenhandedly among all. It is an agreeable idea, but nothing like it has ever happened in human history, and there is nothing in the behaviour of nations today that suggests that it is ever going to happen.

The lessons of history suggest that when there aren't enough resources to go round, we fight for them. Animals fight for them, too. But each one fights only for what it needs, at the moment of need, and the shortages in the animal world have been accidents of season rather than a consequence of global asset-stripping. There is another difference. Animals have some very sophisticated weaponry for offence and defence; they also have their own versions of the arms race. But they are not equipped with colossal stockpiles of napalm, nerve gas or thermonuclear warheads. There is — or at least there has hitherto been — room in the biosphere for hummingbirds and whales and elephants and Siberian tigers, because each takes what it needs. Homo sapiens, who came out of the last Ice Age as little more than tool-using omnivores surviving on the margin with the help of flint and fire, could end up back on the margin in the caves of ice, simply because we ignored the lesson of the coal tit and the hummingbird, and abandoned the sensible use of energy.

We could still take up the option. A number of energy and development strategies are already on the table in the world's debating chambers. They have not been devised by those sandalled visionaries contemptuously dismissed as 'deep greens' or 'eco-freaks'. They have been placed on the table by smooth-suited analysts from the capitals of Europe and America, by the international bureaucrats of the United Nations, and by political sages who have earned worldwide respect. Some of the recommendations rest on technologies for saving power: the incandescent light bulb needs 75 watts but the fluorescent one needs 18 watts and lasts 10 times longer: a switch could save billions. If you painted Los Angeles white and planted trees in every garden, you could save 20 per cent on air conditioning bills. Others are guaranteed to provoke alarm. If energy is to be fairly distributed according to populations, then the USA will have to cut its fossil fuel consumption by 85 per cent, claimed a Swedish professor at a world forum.

These are, by and large, no more than statements of principle, but the mechanisms for saving energy and sharing resources and pursuing new solutions are already in place. It would in theory be quite possible to devise and agree a flexible master plan which, with the consent of all, rationed and apportioned energy and resources according to those who needed it most, while those nations which were already comfortable held back from the trough. In the interests of fairness, there might be curious but logical reversals. Tribespeople in the African savannahs could rely on air-conditioned four-wheeled drive vehicles while the well-blessed in Southern England or urban California could use bicycles and public transport. It sounds improbable, but not unreasonable: it would simply be a mild parallel of the mixed energy-strategy solutions of the animal world. It would enable individual groups or regions to use most efficiently the energy, the resources and the techniques to hand, to come to terms with the particular dictates of culture, latitude and soil, and achieve a level of security and comfort. That level will differ from group to group, but we'll have no trouble recognizing it. It will be the one at which each population becomes stable.

But to achieve that will require selflessness on a scale unparalleled in history. There has never been any evidence of sustained selflessness on a national scale. Hitherto, the dispossessed — the desert Arabs, the Goths, the Vikings, the Mongols of Genghis Khan — have had to take what they wanted by force; and the fearful possessors have used whatever means they had to hand to defend their wealth. The history of the last 3000 years has also shown that the rich nations, armed with superior weaponry, more professional forces, and stronger national institutions, have shown a pronounced eagerness to set forth to make themselves even richer at the expense of the poor. The process is still going on. Foreign loans, aid budgets and trade and defence agreements have in effect doubled as weapons of conquest; they have kept the client nations poor and fuelled the economies of the rich.

But things have changed. The means of offence and defence have become more effective in the twentieth century, but they have also now begun the process of diffusion downwards to the poor. These weapons have their own impact on climate. In a cosmic felicity, while we learned the lesson of the greenhouse effect by observing Venus, the planet named after the goddess of love, we learned quite a different climatic lesson from the study of the frozen, lifeless planet Mars, named after the Roman god of war. We learned what might happen to a world after a full-scale thermonuclear war. It would indeed be the ultimate irony if humanity were to reverse the greenhouse effect by resorting to open warfare for resources — thus triggering world-wide conflagration for the right of the few to drink champagne and eat stuffed goose livers, while millions begged for rice or bread or cornmeal in the streets — and cancelled global warming with a nuclear winter.

CHAPTER 6

The coming of the big heat

At several points in 1988 and 1989 it began to seem as though an invisible sheet of glass had been stretched over the northern hemisphere. In July of 1988 two senior government scientists in North America — each at a level more concerned with crisis management than with science, more concerned with kicking governments into action than with rainfall statistics and temperature averages — caused a flurry of international alarm and a furore among their peers by announcing that the greenhouse effect had begun. Among the most irritated were those scientists from the universities and pressure groups who had been warning — in some cases for decades — that the greenhouse effect was about to begin. This is, of course, because scientists are what they are: science is about knowing, not guessing. And the greenhouse effect is going to be one of those things which we will know from hindsight. Twenty years after it begins we will be able to look back and say, 'Ah, clearly it began then, 20 years ago.'

The problem is that annual temperatures have always fluctuated. The fluctuations can last for decades. Few people took Arrhenius seriously in the nineteenth century. Few took the British scientist G S Callendar seriously 50 years ago when he examined the figures and made substantially the same predictions about the impact of fossil fuel burning on the world's carbon dioxide levels and the atmosphere's average temperature. That is because instead of getting warmer in the post war years, as he suggested, the world, for a while at least, actually cooled.

The other signs, too, are not easy to read. Coastlines somewhere are always being washed away, or flooded. Some landmasses are sinking: others, unnaturally compressed for

The radiation of the sun and the greenhouse gases

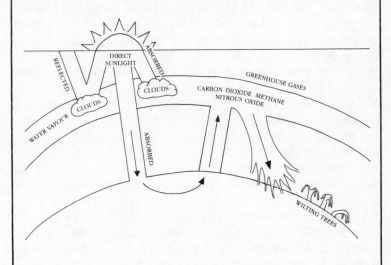

The visible light from the sun hammers the earth every day. Some of it is reflected into space from the clouds (and the ice-caps too). Some of it warms the upper atmosphere. But much travels through the atmosphere as if it was glass and falls on the rocks and the soil and the vegetation and the oceans and is absorbed, warming the surface of the earth. Although the energy arrives as visible light, it is re-radiated back into space as infra-red wavelengths as the rocks and oceans cool. Greenhouse gases let visible light pass through them on the way in, but trap the infra-red on the way back. If there were no greenhouse gases, the world would be very cold. But as these gases increase because of fossil fuel burning and forest destruction, the global temperatures will increase.

100,000 years because of the weight of glaciation during the Ice Age, began rising 10,000 years ago and are still imperceptibly inching upwards. The hurricane that shook southeastern England in 1987 was the nation's worst storm in 300 years, but it can't be blamed on the greenhouse effect, because of course there must have been a similar storm 300 years ago in the days before the Industrial Revolution. Unseasonably warm winters are, by definition, unusual, but there is a long historic record of them. The drought that hit the great plains of the United States in 1988 was the trigger for the first alarm signals but it wasn't evidence. There were droughts and dustbowls in the 1930s — John Steinbeck used the dust and the pitiless skies as the backcloth to his novel *The Grapes of Wrath* — and again in the 1950s. (In the Depression Years, folk wisdom blamed the newfangled radio waves for the dustbowl; after the Second World War wiseacres blamed the atomic bomb tests. Scientists with long memories have every incentive to hedge their bets.)

So, for the time being, all phenomena remain only portents. We will recognize the greenhouse effect by checking what happens against the predictions. These predictions are based on mathematical models. The science of modelling the future is (who needs to be told this?) somewhat less than perfect. The climatologists are, furthermore, basing their models on data from an atmosphere and ocean which they don't understand particularly well, and they are modelling processes which have chaos built in. The mathematicians who devised the models know that they are shaky; other people standing by and watching might wonder whether it mightn't be simpler to study the entrails of a goat for the answer. Even so, a number of calculations have been made. The world has between 500,000 and one million kilometres of coastline (why this figure should be so imprecise is because of fractal mathematics: do you measure the coast of China as a big smooth curve, or do you include every bay and inlet and estuary, and do you allow for every rocky serration or sandbank within those bays, inlets and estuaries?) Most of the world's great cities are on the coasts. There are huge concentrations of people on the great river estuaries. A 1 metre rise in sea levels could affect land all the way up to the 5 metre contour. One calculation is that it could thus affect a total of 5 million square kilometres, including one third of the world's cropland and the homes of one billion people.

The word 'affect' is imprecise. This is because different terrains throw up different impacts. In a storm surge, beach and dune is at risk for 60 metres or more. In a riverine estuary, a storm surge at the right tide could sweep tons of saltwater over riverside cropland for 50 miles or more. Even now, the Rhone is salt 30 miles from its mouth when the conditions are right. In Shanghai a typhoon once pushed the sea up 5.74 metres above normal. Shanghai is home to millions; its countryside, drained by the Yangtse, feeds and clothes hundreds of millions. The flood plain of the Yellow River — it is also known as China's Sorrow — is actually in some places 20 metres below the river. Storms and floods blight these regions now: when the sea is a metre higher they may be untenable. In China, our oldest civilization, where there are 1.1 billion mouths to feed and where only 10 per cent of the land surface is suitable for agriculture, land is being lost when and where we can least afford it.

But the sea's intrusion could 'back up' the ground water to flood farms and rot the basements of cities; its pounding waves, climbing ever higher, could accelerate cliff erosion, sweep away poulders, levees and dykes, wash over and poison reclaimed land and destroy the spawning grounds of fish. It could also wreck local, and distort national, economies. The Dutch, who know about the anger of the seas, calculate that they will have to spend an extra $5 billion by 2040 to save the land they have reclaimed so far. They must. It is home to eight million people. But there is reclaimed land to lose in America too. Miami is an artificial city, with an artificial beach, built on what was once everglades swampland. It exists as a holiday resort, a place in the sun: the slums, the refugees and the criminal underworld are an afterthought. But the beach is at risk now from Atlantic storms; Miami, too, could crumble — at least economically — before the advancing tides. The groundwater is only feet beneath the city streets. Fresh water floats on salt: rising sea levels could push the water in the Biscayne aquifer up another metre to hammer roads and warp bridges and convert the parks back into swampland.

The low-lying 'barrier' islands that protect low-lying coasts, in the Wadden Sea off Holland and Germany, off Texas, of the mouths of the Mississippi are all at hazard. So too are the coral islands. The Maldives in the Indian Ocean, with 1190 atolls, and the people of Kiribati and Tuvalu (once the Gilbert and Ellice Islands) in the Pacific, fear for their future. There has already been an international conference to sound the alarm and try to pave the way for a solution. Coral reefs are the creation of the seas. It was Charles Darwin who first conjectured that the fringing reefs grew as the seamounts sank. Seamounts are old volcanoes gradually subsiding as they scud slowly on the spreading ocean floor away from the fracture zones that created them, and as they sink, the corals, tiny creatures, grow, flourish and die in the warm, sunlit shallow waters of the tropics. The reefs are among the most diverse, the richest habitats in creation, but they are a patina of life renewed annually upon its own necropolis, for the coral limestone which is the bedrock of the atolls is only the impacted bones of aeons of living coral. Thus such islands can never be high: 2 metres above sea level, sometimes 3. A rise of 1 metre is enough to make life untenable for humans on many of them; they are on the margins of human habitation now, and rising oceans could mean that the disastrous once-in-a-lifetime storms could return once every few months.

Nor is there any likelihood that the uninhabited, fringing reefs that protect many islands from the ocean's implacable storm waves could survive. Some corals may grow swiftly enough to keep pace with rising sea levels. If they do, they will be the branching corals, the ones that break off in storms. The others will gradually be submerged, their growth suppressed a little more simply by their growing distance from the surface. The atolls were never paradise — they support coconuts and pandanus for matting and nuts but little other plant life, they have no fresh water except rainfall, they may be 20 miles long but they are not usually more than half a mile wide, and they are pounded by steady 20 to 40 miles per hour winds most of the time and hurricanes and typhoons interrupt development, trade and supply — but they remain the dream islands of the grey, urbanized West. When they go, there will be less to dream about, and hundreds of thousands from the Indian Ocean and the Pacific will have to move; to find new and stifling homes in the crowded cities and hostile cultures of the Pacific basin.

But there are other, grimmer predictions. Guyana is a small nation, with a population of less than one million; a sea level rise of a metre will affect about 3 per cent of the nation's available land. But that 3 per cent houses 90 per cent of the population. Tragedy is likely to overtake Egypt. It is a country composed of history and sand. Only 4 per cent of its soil can be cultivated. Its life is centred on the Nile. Photographs of Egypt from space, taken at night, show a large blank space, with threads of light following only the course of its river, and fanning out into myriad pinpoints at the delta. But the Nile Delta is sinking. The very dams, built with foreign aid and loans that have yet to be repaid, and which were to store water and provide life and hope for a hungry nation, are trapping the silt that, for millennia, flooded annually over the fertile delta of the Nile, pushing it out a little further into the Mediterranean, raising the soil level with each inundation. Without its annual topdressing, the delta floodplain has begun to settle, and to be washed away at its edges by the sea. And then add to this a rise in sea levels of up to a metre. There have been several studies which attempt to predict the likely results. Put brutally, if sea levels rise by 1 metre, 12 to 15 per cent of the nation's arable land will be flooded. About eight million people will lose their homes. There are brackish lakes along the coast which provide up to 50 per cent of the country's fish catch. They too could be lost. The country will lose about 15 per cent of its gross national product. And Egypt is one of the fastest growing populations in the globe.

There are delta problems everywhere: the mouth of the Po and the lagoon that shields Venice, the Gulf of Cadiz, the north Aegean. The Mississippi region is subsiding at about 2 millimetres a year, more in some places; the offshore barrier islands that protect its mouth are migrating inland, and shrinking fast. The rich can cope: they can throw money into barriers and their flexible economies can cushion the shock for the dispossessed. Bangladesh, once described by an American statesman as a 'basket case', cannot. Bangladesh is a nation on the edge. About 80 per cent of the country is a flattish shelf, built up by the flood plains of the Ganges, Brahmaputra and Meghnas rivers, and one fifth of that shelf is actually water. Many of the farmers of Bangladesh have to cope with floods almost every year, either from the rivers, or the oceans, or both; others face annual drought, and struggle to find water for their crops. Half the nation is no more than 5 metres above sea level. What happens to Bangladesh depends on what the Bangladeshis do in the next 20 or 30 years. If they dam or divert rivers and carry on sinking boreholes then the land is likely to subside at the same time as the sea rises. The choice is a bitter one. The farmers at risk from flooding can lose 30 per cent of their crops and the farmers who cannot rely on flooding or even rainfall may lose even more.

River control by diversion, and irrigation by well and reservoir, sound like the only sensible answers. But in that case, they will suffer the same fate as the farmers on the Nile. Some flooding is always necessary to maintain a floodplain: that's why they exist, and why they are so fertile, because they are built on moisture and fresh silt. But the flooding in Bangladesh has, in recent decades, been catastrophic: it is also our doing; another price that humanity is paying for the clearing of the forests. But that is the subject of another chapter. And anyway, whatever the sufferings of Bangladesh now, they will be tenfold under global warming. One scenario suggests that 16 per cent of the nation's land could be lost by 2050, and before then salt water will be seeping into wells and reservoirs that are now far from the sea.

There is yet another problem. If seas are higher and climates warmer, tropical storms will be more frequent and more destructive. In Bangladesh, storm surges can rise 6 metres above normal and travel 200 kilometres inland. In 1970 surge waters washed over 35 per cent of Bangladesh's land area and 300,000 died, either from the flooding, or the famine that followed. In 1987, 1.5 million people lost their homes in a flood. In 1988 the waters covered 80 per cent of the country, causing loss, anguish and distress to 25 million people and drowning 1200.

I have said before that figures lose their meaning as we add noughts to them. We can feel for the sufferings of the few, because we can make the leap of the imagination that permits us to set ourselves in such circumstances. As the numbers grow, the capacity for empathy fades. We can grasp the enormity of the Nazi death camps of the Second World War if we know about Ann Frank, or Primo Levi, or Eli Wiesel, or the wretchedness of Stalin's gulags if we read of Ivan Denisovitch. The murder of six million is another matter. If we think of it, we feel shock, but somehow we cannot think of it.

This is a paradox. When the few — in tens, or hundreds — suffer, the reaction is to help. Around those few are millions who can, in a great rush of sympathy and generosity, offer money, food, living space, medical supplies, shelter, the embrace of kindness, the consoling whispers of sorrow and the murmuring words of hope. But if the agony affects millions, we can only shrug. But the millions cannot turn to each other for help. The family shivering on their precarious rooftop as food, animals, tools and stored crops are swept away cannot turn to their neighbours on the next roof. All they can see about them is starvation, hunger, loss and despair. The muddy grey waters stretch on every side to the horizons and beyond. We have become dulled to the scale of the tragedies that overwhelm Bangladesh almost annually. But the cycle of flooding and storms is in the end sure to increase: finally, the very gloomiest calculation has it, flooding could mean that by the end of the twenty-first century 38 million people will have to leave their homes and their farms. Bangladesh now has a population of about 110 million.

This will be another signal of the greenhouse effect: the frequency of what climatologists call 'extreme events'. These are events that don't happen very often, random accumulations of circumstances, like those accidental, surprise waves that suddenly sweep anglers away on seemingly safe, placid beaches. Scientists have devised a paper hypothesis. It is no more than that: it can only be tested by time. It proposes that tiny changes in climate can dramatically increase the occasions of disaster. Under Britain's present rainfall regime very severe drought may occur only once a century. Cut the rainfall by only 10 centimetres and such a drought could occur every seven years. Even worse, with even small changes in climate, the odds against very rare events happening twice in successive years fall dramatically. If a drought (or a flood, or a cyclone, or a heat wave, or summer hail at harvest time) tends to occur once every 20 years, then the chances of it occurring in two successive years is very small. Such a thing might happen only once every 400 years. (In theory, it may never happen, but in weather, as in every other aspect of life, Murphy's Law operates: whatever can go wrong, will.) But with a tiny change in climate, a double disaster could go from being highly improbable to probable: say, once every 16 years. Most agricultural communities can absorb one disaster. How many can cope with two?

But if the poor will suffer, so will the rich. The Canadians, who (along with the Scandinavian nations) have taken the coming climatic crisis more seriously than most, have already embarked on a series of studies of what might happen, not at a general level, but at specific sites. They took a cool look at Charlottetown on Prince Edward Island. They concluded that, if sea levels rose a metre, the city could be in danger of losing its main marine terminal and coastguard docks. Costly new developments on the waterfront — including a convention centre and the courthouse — would be flooded at high tide. Several streets within the city would be below high tide level, or flooded during storms. The main storm sewer system would be threatened: water would 'back up' in it at high tides, causing flooding well away from the river. Two sewage pumping stations would be below the high tide mark; the sewage treatment plant might cease to function; 225 buildings away from the water's edge would be flooded and uninhabitable; the town would lose a park, and the government car park and yacht club would have to go elsewhere. Roads, too, linking the town with nearby rural communities would be cut, and an industrial estate outside the town would be at risk in severe storms.

All communities survive on a belief in stability; all investment is a gamble on time with short odds. Farmers each year bet that because last year they could grow wheat and potatoes and crop plums, they will be able to do so again next year, and the year after. The same knowledge governs the choice of barns and purchase of machinery, the decision to employ labour and the ability to spend on their children. They know too that there will be late frosts, spring droughts and summer rainstorms that will damage crops and reduce yields, but not so many that they cannot survive. Their compatriots in the towns build their ports on river mouths because they know that, except in exceptional weather, ships can shelter in them, they build hydro-electric power stations on lakes and rivers because they know that, in the decades ahead there will always be enough water to keep the dynamos spinning, and in the mountains they open ski resorts and holiday cabins because they can turn the cold and the snow to their advantage and convert a precarious upland existence into a busy and secure one.

But the greenhouse effect will change all that. What studies there have been spell stress and distress for communities that have built up a reliance on what has been, and who have invested even more money and hope on the assumption that what has been will be so again, indefinitely. The snow deserted the Alps and the ski resorts of Europe were badly hit in the winter of 1988, while people who fled to the Mediterranean for the traditional escape from the North European winter were startled to find forest fires raging in France, Tuscany and Sardinia. And the great drought of the summer of 1988 in the North American grainbelt may not have been one of those statistical accidents of the heavens that have afflicted farming for millennia. It may have been — there is no way of knowing for sure — a consequence of global warming. The overall gradual heating of the globe may have forced an earlier melting of the snow on the Great Plains. This would have reduced the albedo, and thus increased the warming of the soil, which in turn would have stepped up the evaporation of moisture from the soil too early in the summer, so that the parching began not as the grain ripened, but as the stalks struggled to reach flowering height.

North America is the world's breadbasket. The 1988 drought practically halved the world's food reserves; still worse, there was a similar drought in China which shattered Beijing's hopes of strengthening a precarious economy. The predictions for the future of the grainbelt are not good. There will be higher temperatures and less soil moisture. Some of the Great Plains will become semi-arid rangeland; other parts will continue to grow drought-tolerant wheat instead of corn with twice the yield from the same area; other parts will tip over into desert. There will be a massive shift northward for agriculture. Canada may literally reap the benefit, but there will be a gap. Farming is a slow business, in which time is the first and biggest investment. The farmer does not put his farm under a crop he has never been able to grow before until he sees his neighbour doing it, and succeeding. Southern England may very well become ripe for peaches and sunflowers and aniseed, and Lancashire a landscape of wheat and vineyards, but it won't happen because university-based climatologists say so. It will happen because for years the traditional crops have been delivering steadily lower and lower yields, and the switch will be marked by a growing number of farming bankruptcies. One of the problems is that many crops are sensitive to heat. If it is too hot in those days in July when maize pollinates itself, yields fall. Rice, too, can become sterile under the blaze of the sun. As every farmer knows, it isn't the average heat or the average hours of sunshine or average rainfall that matter so much as things like the first frost, the last frost, the weight of rainfall at the wrong moment, the waterlogging of the ground, the dry spells at the wrong time of the year and the high winds just before the harvest. And all these, too, are another set of imponderables in the greenhouse equation.

An average rise in temperatures can be looked at as an extra degree of heat per day for the year. In effect, it is an extra 365 degree-days thrown onto the climatic roulette table for farmers. But it doesn't mean that each day will be a degree warmer. If they got the lot over six weeks in early summer the consequences could be disastrous. If they got the lot in 10 days they could be staring at a dead landscape. The impact will be even harsher in exactly those communities whose prosperity is built not so much on the richness of their produce as upon the precise flavour and bouquet that happy accidents of geography, climate, soil, water and tradition have combined to create, and the cachet that time and reputation have conferred upon it. Who will want Champagne when the conditions that go to make the world's most famous sparkling wine what it is have been shifted to the fields around Hamburg or Riga? What price Roquefort and Camembert and Burgundy and Bordeaux under skies more appropriate to Algiers?

There are tricky times ahead. They would not be quite so tricky if humankind had not been throwing other pollutants into the air besides carbon dioxide, and if we had not been wasting and fouling the soil and the forests and the water as if more were, so to speak, on tap, whenever we should need them, and if, under the pressure of population growth and the race for development in those nations that have sickened of poverty and hunger, we were not going to go on making those same mistakes again, on an even more massive scale.

BOOK II
The poison cloud

CHAPTER 7

The assault on the skies

Let us pause, and draw breath. When we do so, we breathe in not just the air's oxygen, nitrogen and noble gases but also, depending on quite where we are, oxides of sulphur, ozone, lead, cadmium, chlorine, silicon dust, petrochemical solvents, nitrogen dioxide, nitric oxide, carbon monoxide, dioxins, polychlorinated biphenyls, ammonia, radon, formaldehyde, halogens, asbestos fibres, nicotine, aerosols, mercury fungicides and soot. We are now among the pollutions we know best, the ones we can both live with and deplore. We have names for them: we call them acid rain, which is a catchall for a number of processes, and photochemical smog, which is sometimes more precise but which also includes the agents of acid rain, and toxic metal pollution, which is simple and accurate.

The scale of the assault on the atmosphere is staggering: we pump 25,000 tons of arsenic in tiny particles each year into the air alone (we release six times as much onto the land and into the water); and then the figures go on and leave us reeling: 12,000 tons of cadmium, 6000 tons of mercury, 65,000 tons of manganese, 140,000 tons of vanadium and 376 tons of lead and around 180 million tons of sulphur dioxide. We poison our hearts and lungs, we give ourselves bronchitis, we alter the enzymes in our blood and make our children hyperactive, we induce anaemia and we reduce the capacity of our blood to carry oxygen. And we give ourselves cancer. We do these things by burning coal, making and driving motorcars, mining and quarrying, smelting metals, erecting buildings, and then painting them, and then making petrochemically based industrial products inside them, by growing crops and spraying them against infection by rusts and mildews and insect pests, and then spraying them again to keep the weeds down, and then by using toxic chemicals to maintain the electrical industries on which our leisure is based. We put up with all this pollution in the name of industry, of urban life, of development, of what we now believe to be civilization. And the consequence is that many of us live in cities which can seriously damage our health.

This is not an overstatement. According to a United Nations Study in 1988, roughly 625 million people are exposed to hazardous levels of sulphur dioxide and roughly one billion people have excessive quantities of suspended particulate matter in the air they breathe. Once again, the poor suffer rather more than the rich: although London (until the early 1960s) Los Angeles and Tokyo are famous for their urban smog, the crowded cities of the hungry world are far worse. All their problems are exacerbated by the headlong flight from the land to a dubious promise of hope in the urban shantytowns: quite literally, millions have traded rural poverty for lung disease. Mexico City is the most notorious, and there the air is so bad that embassies routinely warn the wives of foreign diplomats not to have children during their tour of duty. Parts of Eastern Europe, too, have been designated as ecological disaster zones; in some places in Poland the cadmium is 200 times the natural level; in some places in the same country the lead in vegetables and fruit from the market gardens has risen to 230 milligrams per kilo. It is quite dangerous for an adult to consume more than 3 milligrams of lead per week.

We can, up to a point, live with these things: manifestly, we do live with them, and in the period of the greatest expansion of pollution, roughly the 25 years that followed the end of the Second World War, when the skies darkened with filth and the fields hummed with invisible toxins, and whole landscapes were torn up to create new industries and bigger cities, we flourished. Despite the assaults on our health from all sides, the world's population went up from two and a half billion to five billion. At the same time, infant mortality figures began to fall and life expectancy to soar. When, more than a decade ago, the famous 'dark satanic mills' of the English North and Midlands began to close down forever, there was wistfulness, resentment and regret. The smoke had meant wealth and security for millions, and the cleaner air that began to blow about the depressed and forsaken cities had within it the taste not of health but of sterility. A similar bitterness, and even despair, overtook the great industrial cities of the American North, as old industries collapsed and new industries flourished elsewhere. We talk about 'the environment' as if it were an ideal, something that had always been there, and we have spoiled, or might spoil further. In fact, the environment is the world about us, the one we know, the one we helped make, the one we grew up in, and when it changes, even for the better, we miss it a little, soot and scrubbed steps and all.

However in the 1960s there began to be signs that in fact the recent decades of massive industrial and chemical activity were beginning to cause lasting harm. The most obvious, ubiquitous affliction — the soot and sulphur that caked whole streets and corroded the ancient stones and blackened the trees and set whole streets coughing — was not necessarily the cruellest nor the most unnatural. The second-rate coal that kept British home fires burning even before the industrial revolution was already causing complaints: Dickens (and Hollywood and Ealing films about Jack the Ripper) made the 'London particular' — the Victorian smog — a part of the iconography of the British capital. But the soot and sulphur is a manifestation of the best known and up till now the most costly environmental insult. The Organization for Economic Co-operation and Development has hazarded a guess that the complex of polluting events known as acid rain could be costing Europe as much as £44 billion a year. Acid rain is the name given to the shower of weak sulphuric and nitric acids which falls from the skies all over the industrialized world, and downwind of it. The acids are formed by the upthrust of sulphur and nitrogen pollutants from power stations, smelters and motor vehicle exhausts. These soar into the skies and react in the clouds to form acids, and fall again upon the cities and forests and the farmland.

It is worth putting it another way. The soil, the rocks, the hewn stones, the plants, the bricks and tiles and concrete and asphalt, and the people, are being bathed in acid. The acid doesn't have to form as liquid in the skies, of course. The particles of sulphur dioxide can fall to the ground as dust to react with the water on the surface. This acid is weaker, much weaker, than the acid into which certain notorious murderers plunged their victims to destroy the last vestiges of the evidence, but of course murderers only had to do that sort of thing once or twice. The acid rain falls somewhere every day. This watery, corroding downpour has been blamed for the deaths of salmon and trout in Norway, and the sterilization of upland lakes all over northern Europe. It has stunted the development of frogs and toads and it affects the insect life of the waters and thus directly or indirectly the birds that live upon them: the kingfisher, the flycatcher, the dipper. It has leached metals from the soil, and filled the lakes with higher concentrations of cadmium, mercury, iron and zinc and aluminium.

It has also been blamed for the widespread sickness and decline of Europe's forests and woodlands, by gradually altering the chemical balance of the soils and affecting the nutrients within them, as well as by damaging the leaf tissues to slow or stunt their growth. This is still a matter for argument: ozone, the poisonous version of oxygen, which makes the eyes sting in smog from automobile exhausts, has also been blamed. Most botanists think that acid rain is just one of a whole complex of stresses that may be weakening the trees, and that what actually kills them is what has always killed them: drought, parasites and diseases, metallic poisoning and bad forest management. Nor is acid rain always, and universally, a bad thing. Some soils are actually too alkaline for certain plant life: a certain amount of acid rain alters the soil chemistry in favour of growth.

The most dramatic effect of acid rain is upon buildings and statuary. Ancient Rome weathered fairly well for nearly 2000 years until the motor car choked modern Rome. The destruction of the Eternal City's ancient monuments by acid rain in the last 40 years has been about equal to that of the previous 2000. The decorative gargoyles on every cathedral in Europe have been defaced: quite literally. The Taj Mahal in India is being blistered and disfigured by acid rain; Venice and Athens are blighted by it, and some of the stones in St Paul's in London have lost two centimetres in 250 years of corrosion.

But these things are simply natural processes which are now happening at greater speed. 'The Rockies may crumble, Gibraltar may tumble. They're only made of clay' we once sang. Nothing is here to stay. Were there no humans and no industry, weathering would happen, and limestone would still be etched and pitted and stripped by rain which would be weakly acid, and other rocks smeared from time to time with soot. The particles of carbon that form soot are the by-product of fire, and fire is part of nature's machinery of renewal. Humans use it, but they didn't invent it, and fire would be blazing across grasslands and forests without them. Sulphur dioxide and even the lethal gas hydrogen sulphide which smells like rotten eggs bubble from the ground in the thermal regions of the world. The loyal townsfolk of Rotorua, New Zealand, used to be fond of saying that they didn't like the air away from their city: it had, they would argue 'no taste'. The world's volcanoes belch millions of tons of sulphur into the atmosphere every year: only recently has industrial output overtaken nature's. What is happening is what has always happened, only worse.

The same is true of many other pollutant processes: all metals exist in the soil in small quantities or at trace levels (we can only mine or quarry them when they exist at unusual concentrations) and although many of them are toxic in quantity, some of those are also necessary for life and well-being in trace form. The same people who worry about the steady dusting of zinc on the soil go out and buy it in tablet form as a dietary supplement. This irony is easily squared. We need minerals in forms in which we can metabolize them. They also exist in forms which neither we nor plants can take up. And they exist in forms which are destructive. We don't, however, need a great deal of lead or mercury or arsenic in any form.

But we have always lived with poisons, and we have always had our chemical industries. When we store grain for the winter we risk the build-up of aflatoxins from natural moulds: aflatoxins could be the basis of modern biological weapons systems. When we make bread and leave it to rise, or curdle milk to make yoghurt, or set it aside to ripen, with the help of certain moulds, into cheese, we are using biochemistry; and when we bake the loaf over the simplest fire of sticks and leaves we release poisonous carbon monoxide and a shower of carcinogens as well as the greenhouse gases, water vapour and carbon dioxide. All human activity involves chemistry, and all chemistry has within it the elements of Faustian bargain, a pact of sorts with the devil. It helps to bear this in mind. We don't, in the normal way, poison ourselves with bread and cheese because our traditions have established a series of safety regimes and protocols of which we are no longer conscious. 'Don't eat that!' we say. 'It's mouldy. It's gone off!' Very early on in our civilization, we learned to preserve liquids for storage by encouraging the build-up of a sterilizing toxin within them: alcohol is a poison, and even lethal if abused, but the Babylonians and the Egyptians knew that beer was safer than foul water, and occasional befuddlement followed by a headache much jollier than cholera.

We live, sometimes very well, with the risk of what might be called domestic chemistry because we are always aware of the risk. The real joke is that it is some of those chemicals which we adopted because we believed them to be 'safe' that have done the most harm. We have used them liberally because we thought they could do only good, and they have left their mark on the globe most cruelly. We believed them to be safe because they were stable, that is, nonreactive: ironically we believed at least one of them to be safe because it was tested on millions of humans.

DDT, the organochlorine pesticide (organochlorine means that it is a basic organic chemical such as methane or benzene to which chlorine atoms have been attached) was used to spray wells in the tropics during the Second World War, and behold: the troops ceased to die of malaria. Refugees, soldiers and prisoners were dusted with it to rid them of lice, and behold: there were no typhoid epidemics. Crops were sprayed with DDT, and the even more toxic aldrin, dieldrin and other organochlorine and organophosphorus compounds, and, for the first time, chronic overproduction began to become a national agricultural problem. For a while, world food production began to keep pace with population growth, and population growth was itself enhanced by the spread of the new chemical pesticides which limited human disease-carrying organisms. It was some time before the world began to notice that there were occasional terrible accidents with incautious use, and outbreaks of long-term, puzzling sickness and sometimes wretched death among agricultural and chemical industry workers. At about that time, too, biologists began to notice that the birds had stopped singing. (The story of all this is well known, and much better told, in that classic of the ecological movement, *Silent Spring*, by Rachel Carson. It was first published in 1958 and has never been long out of print.)

What had happened was that man had begun to devise and use chemicals that were virtually indestructible. Many of the chlorinated hydrocarbons and organic phosphates are stable: that is, they do not react with air or water, nor break down in the bloodstream. Many of them act by damaging the nervous system: in fact they are the starting point for military nerve gas weapons systems. They do not dissolve in water, but they do dissolve in fat. A predator at the top of the food chain — an otter, a fox, a seal or a falcon — can end up storing ever greater amounts of poison as it devours the smaller creatures which have themselves eaten the insects that first carried the poison away from plant or soil. The larger the animal, the better the chance of survival, and the greater the accretion of poison in the fat, where, for a while at least, it did no obvious harm.

But fat is meant to be used: at times of stress — while surviving hardship, while giving birth or suckling or rearing young, while migrating — the fat is released into the bloodstream as emergency rations of energy, and the toxins go with them. The adult may survive, but cub or pup or nestling may not. And whether or not either of them survives, the lethal chemicals remain, to find their way eventually back into the environment, as the carcasses rot, or are consumed by scavengers, or slowly decompose in water or the soil by bacterial action. The scale of destruction was massive: some creatures slid almost to the edge of extinction before many of the pesticides were withdrawn and their use prohibited or very tightly controlled. Even so, DDT is everywhere: diffused, it now circulates the entire globe.

So too do the polychlorinated biphenyls, better known as PCBs. These are compounds made in relatively small quantities: they too are potentially lethal. The difference is that industry has always known this: although they were used in paints and carbonless paper and electrical flexes, their chief purpose has always been as a nonconducting insulant for electrical transformers. They are very good for this. They do not conduct electricity and they are completely non-flammable. They are handled with care, and they are supposed to be destroyed when they are no longer needed, either by deep burial or by incineration in special plants. Burial is not a satisfactory form of disposal for compounds that seem to endure forever. Groundwater will always find them, and groundwater will always flow somewhere. PCBs can be transported by water, but they do not dissolve in it: they, too, however, dissolve and concentrate in fat. Incineration is a difficult form of destruction for something that won't burn. In fact the PCBs have to be heated to around 1200°C to break them up, that is, to snap what are basically two benzene rings with chlorine attached to each, at the link which bonds them. Then they can react with oxygen, that is, burn.

But no chemical reaction is ever complete. Ideally, what emerges from the incinerator chimney should be water vapour, carbon dioxide and free chlorine which can then be trapped by other chemistry. In practice, this never quite happens: some of the compounds escape unaltered and the 'ash' contains incompletely burned PCBs, often more toxic than the original fuel. If the incinerator is 99 per cent efficient and handles 300 tons of PCBs a year, then 3 tons escape back into the environment. Three tons is roughly the amount of PCBs known to be draining into the North Sea from Europe's rivers, leaching from burial and accidental escape and careless dumping, and that quantity is enough to worry biologists, who have yet another sickening component to watch out for in the livers of fish and the blubber of seals, the blood of otters and the eggs of birds. The engineers whose responsibility it is to destroy PCBs without letting them escape into the atmosphere are now regularly confronted with samples of contamination in grass and soil around their furnaces. They protest they are not guilty. They may have some grounds for saying this. Although worldwide use of PCBs is falling fast, concentrations of them are on the increase everywhere. They exist in the air over deserts and forests and Arctic ice, they are building up in lichens in the tundra, and over most cities there are enough in the air to prove harmful, in the long run, to small birds.

PCBs are another example of what we reap when we sow the wind. And yet they are a small group of chemicals. In fact, industry has pumped more than 100,000 chemical products into the environment, and we know the long-term toxicities, breakdown pathways and carcinogenic track records of only a few of them. For the last decade, the investors, scientists and engineers of the chemical industries have been the bugbears of all environmentalists. This is somewhat unfair. Only the comfortable and the secure have time to worry about what might be happening in the invisible assault on the environment, and the comfort and security of the West has rested on an Industrial Revolution which had itself been secured by the courage, knowledge, skill and inventiveness of the physical and organic chemists. Since the days of Sir Humphry Davy and Michael Faraday, chemists have gleefully wrestled with obdurate and sometimes volatile matter, often discerning the possibilities first, and building up the theory much later on, to create electroplate and printing ink, durable paper and household disinfectants, buttons and fibres and pigments and dyes and the acids that etch microchips; preservatives and plastics and the bubbles that foam in lager beers; bombs and solvents and photographic film and pharmaceuticals, and in the course of it, they have changed the world.

The problem is that, whereas we — the ignorant public — know by some process of domestic tradition the hazards of the chemistry of food making and decay, we really know very little about chemists and their chemistry. Our collective unconscious still bases its picture of them on memories of films involving Dr Jekyll and Baron Frankenstein (all enigmatic retorts and pipettes and galvanic apparatus) and of older apprehensions about the medieval alchemy — the dabbling in the magic locked in things — from which it grew. Chemists are, as a class, rather relaxed about compounds like PCBs, rather in the way that mining engineers are offhand about gelignite and electricians nonchalant about high voltage cables. Chemists are liable to point out that there are many worse things in the world than PCBs. But the chemists, too, suffer from the blindnesses that come from standing too close. They may be about the last people to see the calamitous consequences of their own researches. They are, inevitably, inside the laboratory, and an industrial research centre is no place from which to observe the abortion of malformed sea pups in the Baltic because of the PCB levels in the mother's bloodstream.

But PCBs are just another example of something we knew was dangerous but thought could be handled safely; a gamble we took and decided we had lost. The deepest irony is that the most dramatic assault on the atmosphere has been made by a group of chemicals which was devised to be utterly safe, and to improve the lives of billions. And the assault was made in a region of the atmosphere of which, 10 years ago, most people had never heard. And the manner of the assault was so improbable that for some years, many atmospheric chemists found it difficult to believe. We are talking, of course, about chlorofluorocarbons and the hole in the ozone layer.

CHAPTER 8

Lethal light and deadly air

Thomas Midgley, a Pennsylvania chemist who was born in 1889, is the man who thought of introducing tetraethyl lead into petrol to stop engine knock in automobiles. Having made it possible for the car to run efficiently, and begin the process of showering the entire globe with lead, he then turned his attention to the refrigerator. Not many decades ago homes that could afford one kept meat, butter and milk cool in an 'icebox'. It was, literally that: a large insulated cabinet that held a huge block of ice which would slowly melt through the week until the ice-man came again, staggering up the garden path with another huge block between a pair of callipers. The refrigerator existed — big commercial ones were used to make the ice — but it wasn't safe. The coolant fluid was ammonia, and if it leaked, it was dangerous.

Midgley went in search of a nontoxic coolant that would boil somewhere between 0°C and 40°C and arrived at a group of fluorine compounds: a certain amount of tinkering led him to what we now know as the CFCs, and in 1930 he stood up before a scientific meeting, inhaled a lungful of the stuff, and blew it very softly round a burning candle, which promptly went out. He had done it. He had made a refrigerant material with all the correct properties for efficient use, which harmed neither man nor animal nor plant, which would not react with anything, would not burn and which would, if necessary, smother flame. The work he started led to results he could never have imagined. CFCs are now used in refrigeration and air conditioning systems everywhere; they are used as solvents for 'washing' microchips; they are used to 'blow' insulation foams, sometimes for frivolous uses like hamburger boxes, quite often for more serious uses like insulants in refrigeration plants; they are used to drive the sticky foams that children squirt at each other, and they are used to fire some aerosol sprays. A companion group of related gases have found a role in fire fighting. They are used most effectively in those places where water sprinklers or other chemical foams would be as disastrous as fire itself — art galleries, libraries, archives, computer control centres and so on.

Both of Thomas Midgley's chemical pigeons started coming home to roost some time during the 1970s. Research was beginning to demonstrate that children with higher levels of lead in the bloodstream were more likely to have dysfunction of the central nervous system. That is, they were more likely to be irritable, hyperactive, clumsy and slower to grasp ideas. Some of these findings were the subject of much argument, and not necessarily from people who had an interest in preserving the status quo, such as the motor manufacturers, the petrol station suppliers and the firms who made the lead additives.

It is notoriously hard to compare groups of children: in this case it meant trying to compare the behaviour of children who lived in the countryside with children who lived in the inner cities, so that if a difference was observed, it could possibly be accounted for by factors other than lead intake. But over time, the findings were confirmed, and then confirmed again. There was no doubt how the afflicted children were getting the lead. They were breathing it. They were swallowing dust coated with lead as they played on the ground. They were drinking it with their water, they were eating it with vegetables and bread, they were eating it with meat. Some of it came from old lead-based paints and lead water piping. In a few places it came from the chimneys of lead smelters. But most of it came from motorcar exhausts. By the time the British government got round to taking the subject seriously, most other Western nations had begun to take action: urban lead levels in North America had actually begun to fall even before the British accepted that the problem existed at all.

But lead, too is an interlude: a problem created by the powerful nations and which, with a bit of pushing and shoving from the electorates and the scientists, can also be solved by them. Pushed into thinking about alternative, cleaner internal combustion designs, the manufacturers thought about them, and then swiftly began boasting of their ecological consciences. The ozone problem is different. The ozone problem is, to put it idiomatically, a beauty. It requires a solution agreed upon by practically the whole world: rich and poor, East and West, theocracies, democracies, oligarchies, those governments sometimes cheerily referred to as 'kleptocracies', because so much public money sticks to the fingers of the heads of state, and even tyrannies, nations at war with each other, economies in decline and economies expanding. The solution doesn't just have to be agreed upon: it has to be acted upon, and swiftly. And even if it is there won't be a noticeable improvement. That is, the problem will go on getting worse for 50 years or longer.

Let us go out into the air again and look at the sky, and breathe once more. What keeps us alive is the gas oxygen. Its normal form is a molecule: two atoms bonded together. Like hydrogen, it can't survive on its own. We know hydrogen as H_2; likewise we know oxygen as O_2. It can exist simply as O, but very briefly: the single oxygen atom is very reactive, and seeks, so to speak, another atom to which it can cling. Why this should be so is a matter of its nuclear and electron structure. All through the periodic table of the elements there is a rhythmic pattern of reactive elements and inert, or nonreactive ones. Helium, for instance, and argon, are inert. Hydrogen and oxygen and magnesium and potassium, for instance, are reactive. When hydrogen exists simply on its own, as a gas, rather than as a component of water or alcohol or methane or benzene, it exists as a two atom molecule. Only then is it reasonably stable. Oxygen, when found as oxygen, and not a component of water, or rust, or sulphuric acid, or alcohol, is stable in its double form as well.

There is another form of oxygen. It exists as three atoms. This is known as ozone, O_3. Although we talk about going to the seaside and smelling that health-giving ozone, we are talking nonsense. We are smelling salt, dead fish, rotting seaweed. Ozone is poisonous. It exists down here naturally in tiny quantities and for fleeting moments — after lightning, for instance — and in larger quantities mostly as a pollutant, a consequence of a series of unpleasant chemical reactions involving other pollutants. If the concentrations of ozone in the atmosphere are high enough for you to notice them (smarting eyes is an early sign) they are high enough to harm you, although most of us seem to survive most of the time.

But ozone nevertheless, is necessary to our lives. There is no life on Mars — that is, on the surface of Mars. There are no organic compounds from which life could form, or which might have been composed by living forms. There may be some under the soil — meteorites believed to have come from Mars have been found in the Antarctic (where they could not easily be contaminated by Earth life) and they have been found to contain organic material. This does not, of course, mean that life existed, or exists, on Mars: there is a whole class of chemical compounds — the hydrocarbons, such as methane and benzene, the alcohols, some acids based on hydrogen and carbon, halogens such as chloroform and substances such as glycine and urea — which are associated with life but which are not dependent upon it. But it does mean that whatever has kept the Martian surface free not only of life but of organic chemistry must be acting just there, on the surface.

The answer is not at all myterious to planetary scientists. It is not so much what is in the Martian atmosphere as what is missing. It is the light of the sun, pounding the surface of the Red Planet through its thin air, which has stripped Mars of the possibility of life, and sterilized its upper soil, and which with each Martian day prohibits any possible life formation; this, of course, is the same sun that fuels, drives and caresses life on Earth. The difference between the two planets is in the composition of the atmosphere, and for the purposes of this chapter, what matters is that, on Earth, but not on Mars, high up, above the biosphere, above that level at which life normally exists, is a layer of ozone. It is this poisonous form of oxygen which filters out the rays of the sun that are hostile to life, and lets through those that are necessary to it.

We need to go back to the sunlight for a moment. It is, of course, energy in the form of electromagnetic radiation: a continuous multi-billion-years-long flash from a massive hydrogen fusion reactor 93 million miles away from our planet. Almost half of the radiation is in the infra-red: the bit we can't see. About 43 per cent of it is in the spectrum of visible light — simply the range of light that human eyes are accustomed to (although some reptiles can see more at both ends of the spectrum) — and 9 per cent is in the ultraviolet.

The ultraviolet section has the highest energy. All radiation can be thought of as discrete, separable packets of energy, measured in wavelengths. All matter can absorb energy (this is why an iron roof, or for that matter a sunbather or a raisin, feels hot in the sun). But each atom or molecule — here we are again with nuclear and electronic structure — also exists in particular states, and can only absorb particular packets or quanta of energy. If an atom absorbs one, it makes a quantum leap from one energy state to another. The reverse can happen. An atom in a very hot body can release energy, falling to a lower quantum state in the process. You can see this happening when a log burns. The combustion of a log with oxygen is actually releasing energy. It is, of course, releasing the energy the sun put in there in the first place to make the wood out of carbon, water, photosynthesis and a handful of minerals. That is why you can warm your hands on it, and read a book in its light at night.

It is easy to understand this when you think of energy absorption as a tree growing, or energy release as a log burning. But what happens to a growing tree or a burning log is simply the sun of a billion tiny molecular energy deals. These exchanges occur at different wavelengths for different elements. Chemists can use these phenomena in a technique called spectroscopy. That is, by studying the wavelengths of light that are emitted from, or which do not pass through, clouds of atoms, they can identify them. They can identify the composition of the outer layers of the sun and the stars through spectroscopes, and they can even guess at the nature of invisible elements drifting through the cosmic blackness of interstellar space by the gaps in the light that reaches us through the blackness; gaps that could only be left because the light at a particular wavelength has been absorbed by hydrogen, or alcohol, or whatever, floating out there in the eternal night. It is the energy traffic of individual molecules that concerns us in the ozone layer.

Think then of the energy of the sun, in the form of radiation packets — quanta, or photons — continuously pounding the upper layer of the atmosphere, where the last wisps of air flail at the emptiness of space. The lower energy wavelengths, the ones we feel as heat, and see as the light which scatters to give a blue rinse to the broad fields of the noon sky, get through. At the very highest level, at up to 300 kilometres over our heads, the x-rays and some of the ultraviolet ones slam straight into the gases and fracture them; dislodge electrons in their outer structure and leave the remaining atom as an ion; electrically charged instead of electrically neutral.

This is the ionosphere. It is the first, the thinnest, the outermost skin of the globe. It is a skin which stops some waves of radiation coming in, and others going out. The scattering of electrically charged particles, fizzing and restructuring, and then being burst asunder again at the margins of the world, is mostly penetrable: we can see through it, we can feel through it, we can launch rockets through it. But the x-rays from the sun and the outermost stars don't get through it, and the same action leaves us with an invisible shell that stops radiowaves from escaping. That is why we can tune into Radio Peking, or Radio Moscow, or Radio Voice of America on the shortwave, because the radio waves are leaving transmitters in Moscow, soaring skywards, bouncing off the ionosphere and being reflected back to receiving sets everywhere on the globe.

Other packets of energy in the ultraviolet penetrate a little further. They don't normally get far. Sooner or later, one of them hits a double atom, a molecule of oxygen. The energy in the packet is just enough to burst the two atoms asunder. O_2 becomes O. But O is an anathema, even to itself. As soon as it meets another molecule of O_2 it conjoins with it. A molecule of ozone, O_3 is born. The process of absorption goes on. Ozone absorbs ultraviolet too, and it stays absorbed. Something like 3 per cent of the sun's rays are blocked in this way, and that 3 per cent includes much of the ultraviolet that gives us skin cancer.

It isn't a once-and-for-all process. It is a continuous one, because the same ozone molecule that swallowed up one particular wavelength of ultraviolet is broken down by another, slightly longer wavelength: broken down to oxygen again. Or it sinks to a lower level, to survive a little longer. In the same way that the soldiers behind the front line are slightly less likely to be struck down by the fusillade from the enemy, simply because the bullets are more likely to be absorbed by those comrades in front. In fact the 'firing line' that is the ozone layer is a deep one.

It is important to remember that although the oxygen-ozone-oxygen-ozone cycle goes on and on, and that essentially nothing is permanently changed, each packet of the ultraviolet radiation is absorbed forever at each interaction. It always reminds me of that passage in Winnie-the-Pooh, in which the world's favourite Teddy Bear is dragged down the stairs by Christopher Robin, with the bear going bump, bump, bump at every step. Finally, there are no more stairs for a bear to bump against. Finally, most of the wavelengths we have reason to fear have been filtered out by the collisions down atmosphere's staircase. The bumping begins somewhere near the edge of space, and extends to about 10 kilometres above our heads. And curiously, it is warmer than the air below it. In theory the temperature should fall as the atmosphere thins in the stratosphere. It does. But at some point it starts to rise again. The molecules of ozone in the stratosphere have absorbed the fiercest blasts of the sun, and held them.

The ozone layer is a shield: an imperfect one, because it is thin. At any time, most of the molecules in the ozone layer are the gases of the air in their normal forms, and thinly spread. If you could somehow 'catch' the ozone layer in a membrane and bring it down to ground level the pressures down here would flatten it to about 3 millimetres. It is imperfect, too, because its thickness varies, not only with time — the creation of ozone is, by its very nature, a daily process — but also with latitude. It is thinnest over the Equatorial regions, and tends to get thicker over the high latitudes and the Poles. It protects Europe and North America better than, say, the peoples of the tropics, who tend to have darker skins to block the ultraviolet that gets through: they need an extra line of defence.

The ozone layer is difficult to measure: it is like one of those sums set for students of the calculus, involving a running tap, a bath and an open plughole, in which the tap is adjusted daily and the width of the plughole is not known, and the body of water in the bath is anyway uncertain from the start. But the ozone screen is effective. It is effective because it is deep. It extends for tens of miles, and it stops most of the rays which have the potency to damage some of the tiniest forms of life, such as the plankton in the waters which are life's first base, and the cells of plants, and the cells of human skin and the retina of the human eye. Mars has no such layer, and that is why Mars is unlikely to be colonized by any form of life with which we could be familiar.

Once again, our jaws should drop. We should marvel. Something — Gaia, God, Nature, the Biosphere, Evolution — has organized the atmosphere so that everything should be just so, just right. Light is good, but not all light is good. Oxygen is vital for us, but one rare form of it is harmful. How perfect that the form of oxygen that is hostile to life should be just the shield we need to protect us from the sunlight that is most dangerous to us, and that this dangerous shield should be in place not where it could possibly harm us, not, for instance at the ceiling to which the condor soars, but much higher, so high that we could not, because of the thinness of the atmosphere, exist at all at that level. And so the heat of battle — this continuous combat between lethal light and deadly air — begins an Everest upon Everest above our heads, and is conducted so silently that only in the last few decades have we even had an inkling of it; how perfect, too, that the ultraviolet light that we should fear most continuously creates its own enemy, which is also our enemy, and that it should all work so precisely that for most of our existence we have neither known about it, nor even needed to know about it.

CHAPTER 9

The hole in the stratosphere

However, some atmospheric chemists almost immediately had an inkling that the ozone layer might be subject to human influences. Thomas Midgley in 1935 suggested that chemists should find a way of actually increasing its thickness: to do so, he argued, would limit still further the amount of ultraviolet getting through, and reduce crop yields the natural way, without benefit of bureaucratic support mechanisms, there being in America both a glut and a depression at the time, forcing some farmers to destroy their crops because they couldn't recoup the cost of harvesting and selling them. Others, a few decades later, were more troubled by the thought that human activity could harm it, and these days there is a serious fear that lower crop yields are more likely to come from a *thinner* ozone layer.

So far we have thought of the atmosphere as a currency and commodity exchange, an enormous, inexhaustible trading floor for oxygen, carbon dioxide, energy and water. But, we must now also think of it as a sky-high, world embracing chemical laboratory, bubbling with hundreds of reactions involving chemicals we don't, in the normal way, need to keep in mind. To try and contemplate more than two or three at once is to induce a headache: to think of half a dozen simultaneously can bring on a migraine. None of the processes of reaction is simple: this is because none of the reactions can be thought of in isolation, not, at least, in terms of their overall effect.

Where three atmospheric chemists are gathered together, there is an argument. Where five are together, there is usually a furious row. Dimethyl sulphide, for instance, rises from the oceans and may make the clouds billow by giving water-

The ozone layer

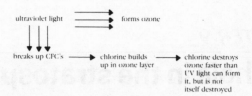

In a reaction that nobody expected, the apparently indestructible CFC's break up under ultraviolet bombardment high in the stratosphere, releasing free chlorine which is damaging the ozone layer faster than ultraviolet light can form it from oxygen. Since the ozone layer blocks the ultraviolet from the lower atmosphere, more is getting through, reacting with industrial and traffic pollution, and creating unwanted ozone in our cities. Although the damage is being done everywhere, the most dramatic destruction occurs in a freak set of geographical conditions each spring, over the Antarctic.

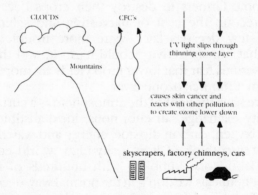

CFC's — the aerosol and refrigerant gases — could account for 14 per cent of global warming (as well as the destruction of the ozone layer).

Low level ozone is produced by traffi and industry, and also by ultraviolet rays penetrating the thinning ozone layer and creating ozone in the lower atmosphere.

vapour something to form droplets around. There is as yet no consensus about what effect this may have on global warming. Methane and sulphur dioxide and hydrogen chloride are all produced by natural processes on the surface of the earth and oceans, and soar upwards to affect the processes going on over our heads. Most of the air is composed of nitrogen: this is essential to plant growth and keeps turning up again in the atmosphere in combinations with oxygen. In all these cases, the gases occur naturally, but our activities have increased their presence. What follows is a description of some of the processes in the destruction of ozone layer. It is, one hopes, confined to the simple, two-aspirin, headache level.

If, for instance, there happens to be nitric oxide about when the sun's more powerful rays split an oxygen molecule into two, then the two single atoms are likely immediately to join with two atoms of nitric oxide to form nitrogen dioxide — and in the process release energy. Sometimes you can see it. The night sky, when there is no moon, has a very faint aura, or airglow: that is, the space between the stars is not quite black. The process blocks the formation of ozone. But that is not the end of it. On top of that, oxides of nitrogen can, acting as catalysts, interact with ultraviolet light to destroy ozone. At the end of this particular ozone destroying action, nitric oxide forms again, and so the whole process can repeat itself.

Sunlight, which formed the ozone in the first place, can also break down ozone. The 'traffic' from ozone to oxygen and back is at its most hectic over the Equator, which gets the sunlight most directly, all year round; the winds on either side of the line, themselves ultimately driven by the sun, also sweep the stratosphere's ozone north and south, which is why the ozone layer gets thicker as the latitudes get higher, and the greatest concentrations of all are over the poles. There is, however, nothing uniform about the process. How much ozone there is at any one point depends on the weather, the season, the latitude and the amount of solar radiation (which varies according to the earth's elliptical path round the sun and according to the sun's own internal cycle). It also depends on how many catalysts — nitric oxide is one, and chlorine another, and both exist naturally — there are in the stratosphere to speed up the reactions.

There is also a certain ambiguity in the reactions (there is at any time more than one chemical process going on up there). There is one reaction, involving a swap of an oxygen atom from chlorine monoxide to nitric oxide, thus creating nitrogen dioxide, which dissociates in visible light and leaves an oxygen atom free again to combine to make more ozone. There is an irony in this. Down here, along the autobahns, the motorways, the interstate highways and in the clogged city thoroughfares, the nitrogen dioxides from car exhausts actually produce huge quantities of ozone just where it can do most damage, and help destroy it just where it can do most good.

The first serious, vocal worries about the fragility of the ozone layer involved the oxides of nitrogen and the testing of nuclear weapons. Nitrogen and oxygen co-exist in the air we breathe, but if the air is hot enough, they react. An exploding nuclear device creates more than enough heat to make this happen. It also creates more than enough thrust to launch hot bursts of nitrogen oxides all the way up to the stratosphere and into the region of the ozone layer. Something like 300 megatons of nuclear devices were detonated in the atmosphere before testing went 'underground' in the early 60s. This may have produced enough nitrogen oxides to thin the ozone layer by about 4 per cent, although, of course, no-one can know for sure.

All anyone can be sure of is that the ozone layer became thicker in the years after nuclear testing ceased, before beginning to fall again. Some of the fallout could still be up there. That's the other thing about the ozone layer: in theory, it is a world apart. The heat generated by all the photochemical reactions miles above our heads creates a phenomenon known as temperature inversion. This stops, to a large extent, the mixing of the air down here with the air up there. It isn't a sealed outer world. The invisible skin between the two layers is always being temporarily punctured by natural phenomena. Rain comes down. Thermals go up. The heat from massive forest fires may occasionally be enough to lob material into the upper sky. Major volcanic eruptions can certainly blast material up there, and once there, it stays for years. So, of course, do the exhausts of military jets and supersonic transports and rockets with a burden bound for orbit in outer space. The pollution it receives, it keeps for a while. If, in that pollution, there are catalysts which affect the ozone layer, then they stay up there for a while and go on affecting it.

Even so, there is an awful lot of upper atmosphere; the puncturing has not been thought to be have been all that frequent, or widespread, and oxides of nitrogen do not last forever, and even if they did they would be washed out of the upper atmosphere in the end. Sunlight is (for our purposes) eternal and oxygen is being produced in huge quantities by plants, and the ozone layer is a permanent ozone-reprocessing plant. It is not hard to see why some perfectly respectable academic scientists took the fears of the first few agitators quite calmly. It wasn't until the early 1970s that the CFC-powered aerosol spraycan — itself a very new invention — joined the ranks of agents suspected of destroying the ozone layer. The alarm came in 1974 from two Californian scientists, Mario Molina and Sherwood Rowlands: they are also closely concerned with greenhouse effect research. CFCs came under suspicion because chlorine is known to destroy ozone without itself being destroyed in the process, and because although CFCs are inert and harmless to humans or, as far as is known, to any creature in the biosphere, and are furthermore virtually indestructible in the biosphere, they do contain chlorine.

There is another fact about the chlorofluorocarbons which was not, at first, realized. They can be destroyed — broken up, altered into something else — by intense ultraviolet light: not the filtered light we have learned to live with down here, but the light that first slams into the earth's tenuous outer atmosphere, up there, at the edge of the ozone shield. And since, down here, at the time, most of the CFCs that have ever been produced were still wafting about in the lower atmosphere, and the rest were still in cans or bottles or refrigerant tubes waiting, inevitably, to be released one day, it seemed even then to those people who chose to worry about it that there was a fair chance that CFCs would get into the upper atmosphere and stay there, and do damage.

The alarm was enough to put the ozone layer and the CFCs on the political agenda. The United States Congress formed a Committee on the Inadvertant Modification Of The Stratosphere. The US space agency Nasa took up the research and calculated that if CFC production was held to 1973 levels, there would be a 6 per cent reduction in the ozone layer, which would in turn let through an extra 12 per cent of ultraviolet. By 1978, in the USA, the use of CFCs in aerosol propellants was banned. And there was still argument, prompted more by the manufacturers than the public, about whether in fact CFCs could actually harm the ozone layer, and if they could, how the harm might be done. CFCs continued to go into propellants in other national industries, and they were still made for industrial purposes, and as refrigerants, and for blowing insulation.

The argument — and the research — went on. Chloro-fluorocarbons were found 19 kilometres into the strato-sphere. Chlorides were found to exist even higher. The people who believed in the CFC theory of ozone destruction argued that this coincided with their expectations. The chemists who worked for the CFC manufacturers argued that it was quite possible that chlorine atoms were not scavenging ozone, but reacting with nitrogen oxide to make chlorine nitrate, thus taking both chlorine and nitrogen oxide out of the combat zone. Thus CFCs might actually be protecting the ozone layer. (This does actually happen, as it turns out — in the sky, as in the oceans, there are many reactions taking place simultaneously — but not in quantities that protect the ozone layer from damage.)

There were other arguments. What proof was there that the ozone layer was thinning? How can you be sure you are measuring it correctly? And Du Pont, the chemical combine which made CFCs under the brand name Freon in the United States, pointed disingenuously to the 'absence of proof of immediate harm' as a reason for doing nothing that would harm business: as if CFCs instantly soared to the ozone layer, instantly destroyed it, and the ultraviolet rays that came through instantly damaged skin cells, and people out in the sun promptly developed malignant melanomas stamped with the trademark Freon.

The ozone and CFC row rumbled on for years. The public, which found the argument very difficult to follow, tempor-arily lost interest. The US space agency began a programme of satellite mapping of the upper atmosphere to try to measure any overall decline in the ozone layer. The United Nations Environment Programme, a special agency created in the early 70s to respond to some of the ecological pressure groups building up everywhere, took up the issue at the end of the decade and began talking about the Vienna Convention for the Protection of the Ozone Layer, which was eventually to be enforced by an instrument called the Montreal Protocol. When this process began there was talk of cutting production of CFCs to 85 per cent of present levels, and in February 1985, a group of Belgian scientists reported that the threat to the ozone layer from CFCs was 'distant'. Within three months, the picture was utterly changed.

It was changed in one of those ironies of circumstance that everybody enjoys. The US space agency Nasa, using a Nimbus satellite, and an enormous science budget, was perfectly placed to measure what was really happening. But satellites produce mind-numbing quantities of data (Nasa has six trillion bytes in store — the equivalent of about 38 million books — and is preparing satellite-based observing systems that will yield information on that scale every few months.)

Such data cannot easily be read by human eyes. There is just too much to absorb. Eyes blur. Patterns become more difficult to discern. Concentration falters. Information on that scale has to be scanned by programmed computer systems. The assumption was that since CFCs, or oxides of nitrogen, were likely to be distributed all over the globe, the thinning would be going on almost everywhere, rather than at one particular place. Further, this would be difficult to see in the short run, since ozone thickness varies somewhat everywhere, at all times and in many cases for natural reasons. There would also be false readings. (Sensitive precision instruments are often temperamental; sometimes even petulant and downright sulky: scientists learn to discount such behaviour. They call the occasional startling or just improbable readings 'artefacts'. Among the most notorious of these were the 'canals' once thought to have been seen on Mars.) Thus for perfectly sensible reasons the computer was told to ignore observations that didn't make sense, such as improbably low counts of ozone thicknesses. Which it did. That is why the alarm bells did not ring first in the USA, where the concern had been greatest, and the effort most energetic.

That, too, is why a British Antarctic Survey scientist called Mr Joe Farman, working in the Antarctic, using old-fashioned equipment, and looking up into the sky from an ice shelf in Halley Bay, first saw the hole in the ozone layer. It is not so much a hole as a massive thinning: a basin, sometimes as big as the USA and as deep as Everest, which forms in the Antarctic spring, which lasts for a few weeks and breaks up and collapses as the sun gets higher, and it is now thought to be the principal site of ozone destruction: the cauldron in which, each year, a huge bite of the earth's protective shield is simply destroyed. In September 1987, at the hole's worst point (its size varies from year to year) in one measured, vertical 5 kilometre column, nineteen-twentieths of all the ozone had gone.

What was going on? The theory of a self-repairing ozone system was that even if ozone was being destroyed, there was still new oxygen to meet the shafts of ultraviolet radiation pounding in from the heavens; the winds, the turbulence, the heat of the reactions would keep the battle lines drawn. The shield might be worn, pushed back, penetrated, but it would still be a shield; just a slightly less efficient one. The overwhelming would not be violent or sudden.

The CFCs are split by ultraviolet as they rise into the stratosphere, decades after their release, possibly riding upwards on such meteorological monsters as the 'stratospheric fountain' over the Arafura Sea north of Australia, thought to carry air currents from the lower atmosphere into the upper. Once split, the CFC dislodges one chlorine atom. This will steal an atom of oxygen from a molecule of ozone to produce chlorine monoxide. This new product will last only until it bumps into a free atom of oxygen; then O_2 will form again, leaving a free atom of chlorine ready to steal another atom of oxygen to form chlorine monoxide, and so the process will go on. The calculation is that a chlorine atom could go on doing that 100,000 times before it dropped out of the ozone system. But that is not the only thing that could happen. The chlorine oxide could join with nitrogen dioxide to form chlorine nitrate and drop out of the system altogether, causing no further damage. Or it could conduct a little chemical square dance with nitric oxide that would very quickly free the chlorine atom and leave the nitric oxide and the ozone exactly as they were: nothing lost, nothing gained. In that sense, nitric oxide plays shadow warrior in the ozone layer. It takes the impact of the chlorine, and frees the ozone. So far, so good. The chlorine atom is free to damage the ozone again, and will go on doing so, but at least two other processes are at work, one of them neutralizing the scavenging; the other taking out the intruder. All this suggested that if the battle was being lost, it was being lost slowly and undramatically.

But the Antarctic phenomenon, which has been checked and rechecked, and then checked backwards in time against exhumed Nasa satellite data, tells a different story. What happens over the south polar region is (very roughly) this. In the peculiar conditions of the dark Antarctic winter, a collossal vortex forms; a region of still, frosty, stratospheric air, a holding tank in the upper sky. This, of course, holds ozone within it. Remember that in polar regions there are greater concentrations of ozone to be destroyed. On the ice crystals of the polar stratospheric clouds, as the temperatures sink to $-90°C$ reactions that some people had thought not possible take place. These reactions remove from the upper atmosphere, within the vortex in the skies, most of the oxides of nitrogen and chlorine nitrate — the chemicals that limit or interfere with ozone destruction. If anything, the reactions actually convert the chemicals into ozone destroying forms. All that is left are chlorine oxides, waiting for the sun to kick them into action.

When the scientists started counting concentrations in the Antarctic they found 500 times more chlorine monoxide than their models had predicted. The Antarctic vortex had, in effect, been turned into an ozone destroying machine, waiting for the switch to be thrown. When the spring returns, the switch is indeed thrown. The sun's malign radiation pours into the cauldron, and suddenly it is as if a continent-sized fist were punched into the ozone layer: the chlorine has the field to itself.

The damage is colossal. After a few weeks, the vortex breaks up, and vast areas of ozone poor air circle briefly over Australia and New Zealand and southern Latin America before the whole structure collapses, and the southern summer proceeds, leaving the ozone layer the poorer. The same thing happens in the Arctic — almost. International teams of scientists using high flying aircraft and a Boeing loaded with computers criss-crossed the polar ice-cap. They found evidence of all the conditions for an ozone implosion over the North Pole. The only thing that seems to save the world from a double collapse, one early in the year in the north, the second late in the year in the south, is the geography of the Arctic.

For some reason, the north polar vortex breaks up much earlier, before the headlong photochemical destruction of the ozone can begin in earnest. It is as if everything were ready. Chlorine is present in abundances 50 times greater than the predictions should allow. The 'reservoir' forms of chlorine are much fewer than models allow. Think of it as an explosion waiting to happen. Gelignite. Shrapnel. Bomb casing. Sprinkled gunpowder. Stacked cans full of petroleum. And a lit twine fuse. And then, at the last moment, it starts to drizzle, and somewhere between the start of the fuse and disaster, the flame sputters out. It is not a matter of complacency: not a case for saying 'That's all right then.' There is no guarantee that future vortices will always break up in time, or that the chemical overloading in the north will not somehow enforce destruction anyway. And — cheerful thought — most of the CFCs that were ever made are still down here, and have yet to spiral into the upper heavens: a vast army of reserves, one day to join the assault on the ozone layer from below, and tip the battle still further in favour of the ultraviolet: the light that has sterilized the surface of Mars, and which, at this moment, even before we know precisely how much the ozone layer has really thinned, and at what rate it is thinning, threatens the comfort and peace of mind and even the lives of millions on Earth.

There is now, at the time of writing, an international instrument called the Montreal Protocol which is designed to control substances that deplete the ozone layer. When it was agreed (in Montreal in 1987) it required a 50 per cent cut in production and consumption of CFCs by 1988. Nineteen months later, an international meeting in Helsinki unanimously agreed to a complete phaseout by the end of the century. The first agreement — and the second promise of the toughest action of all — are the work of the Egyptian-born director of the UN Environment Programme, Dr Mostafa Tolba, and perhaps should be his memorial. (Ironically, he managed to secure the decision he had been aiming for all along by a manoeuvre which looked to this observer more like nimble footwork than mature deliberation. He casually sprang the resolution late in the evening, as delegates were looking at their watches and wondering whether they were going to be in time for a reception by the host government. The next day, one delegate complained bitterly that he had only learned from the BBC World Service, by accident, that he had unanimously agreed to a phaseout of CFCs by the end of the century.)

But agreements by delegates at a United Nations meeting don't save the world: actions by nations united in a purpose just might. Any such phaseout of ozone depleting substances will have to include chemicals like methyl chloroform and carbon tetrachloride as well as companion compounds not covered by the Protocol; and any substitute chemicals devised to replace CFCs will have to be carefully screened for their ozone depleting potential as well. And everybody will have to act upon it. This won't be easy. In most countries the so-called 'frivolous' uses of CFCs are coming swiftly to an end. CFCs are superbly designed for their role as, for instance, refrigerants, and most of the world's cooling systems have been built around them: that is, the design of the plant is dependent on the use of CFCs. The chances of finding another safe refrigerant — and it has to be safe down here, to humans, plants and animals, as well as in the ozone layer — which will allow us to drain the old coolant from the machinery, pack it up for destruction, and pour in a few cans of the new, safe version, are not great. There may have to be massive replacement of equipment in industrial and agricultural cold storage plants and the hunt is on for a satisfactory environment-friendly domestic refrigerator (preferably one which actually saves energy as well, because of the greenhouse effect. Everything comes back to the greenhouse effect, even, as we shall see, the CFCs and the ozone layer).

Once again, it will cost money. This will be a problem for the rich nations but it will also be a fillip to industry. It will, however, be disastrous for the poorer ones, most of them in the tropics, most of them in debt, and most of them relying on refrigeration plants to conserve their food supplies and to stop agricultural produce from rotting before it can be sold to the rich nations for cash. This writer, in less than a year, attended four international conferences on the atmosphere. At each one, he was struck by the remarkable maturity, common sense and concern expressed by the scientists, ambassadors and ministers of the Third World nations. Not one of them said to America or Europe: 'This is your problem, you made it, you sort it out. Furthermore, we are black and don't get skin cancer so easily: why should we give up the refrigerators we have installed on money borrowed from you which we may never be able to pay back because somehow all our dealings with you seem designed to keep us poor? And then borrow even more to equip ourselves with something you say that we must have instead, making our debt burden even bigger and your big corporations even richer?'

But then they didn't need to put it as crudely as that. Although all the delegates from the poorer nations expressed alarm, and threw their collective weight behind the drive to tighten the Protocol, they all made the same point. There would have to be help. The rich nations know this too. Because although the Montreal Protocol, in its inadequate 1987 version, does actually limit CFCs, it allows scope for the developing nations to expand their own production, rationed on a per capita basis. But even a per capita ration would be a disaster if every person used that ration. China has announced that it is building CFC plants, and China intends to have a refrigerator in every household as soon as possible. Should China achieve this ambition, one that the Western world effectively achieved decades ago, then that one nation could practically double the world's output of CFCs. China is a nation that wants to escape from poverty. Accordingly it has engaged on a number of expansion programmes. One of these is a massive increase in the extraction of energy from coal, of which it has plenty. This could mean that by the year 2030, China could be releasing three billion tons of carbon dioxide into the atmosphere. Since the rest of the world is beginning to accept that to limit the greenhouse effect, world emissions ought to be *reduced* to three billion tons, the consequences of such action could be calamitous. So, too, would be an expansion of Chinese CFC production. CFCs not only destroy the ozone layer, they are also greenhouse gases. They are not made in great quantities, but each molecule of CFC is 20,000 times more efficient than a molecule of carbon dioxide at trapping heat. CFCs alone could be responsible for up to 20 per cent of the greenhouse effect. CFCs are not the only things that link the greenhouse effect to the destruction of the ozone layer. The thinning of the ozone layer may have an impact on a greenhouse world as well.

CHAPTER 10

The dark side of the sun

The first lifeforms on earth must have survived underground, or in deep water, because the ultraviolet light that may have assembled the first organic molecules such as the amino acids also has the force to splinter biological molecules, including DNA. The so-called double helix (DNA) is the point at which life begins. It is the information carrier and the replication system of life. It contains within it the programme for survival and multiplication. It is odd to think that the formation of life on earth must have once been a bit like the process that now goes on in the ozone layer: with the energy of the sun's most powerful rays continually forming organic compounds, and then fracturing them when they become bigger and more complex. If that is how it happened, then some of the larger molecules must have survived the haphazard process of formation and destruction, because we are here now, and hoping we are here to stay. We are here because of the eventual development of the higher, photosynthetic, carbon dioxide-consuming plants which began the process of building up the oxygen in the atmosphere, which cooled the world and speeded the rate of soil weathering and made possible the huge, complex machine called the biosphere — screened by an ozone umbrella — in which we survive.

But we are still at risk from ultraviolet. The sun *is* life to us, but it is also a death-ray machine. We survive because most of the most damaging wavelengths don't get through (for the technically minded, the DNA-damaging wavelengths are at the moment virtually blocked by the ozone layer, and the worrying wavelengths discussed here, some of which do get through, are between 290 and 320 billionths of a metre.) Those that do get through can cause both malignant and nonmalignant skin cancer in fair-skinned people, who tend to burn rather than tan when exposed to the sun. Basal cell carcinoma, for instance, is virtually unknown in people of Negro origin, except those who are also albinos. People who live in the higher latitudes are less likely to suffer from it than those nearer the Equator, but the chance of developing it increases with time. The tumours arise in skin repeatedly damaged by the sun; 90 per cent of them occur on the head and the neck, and farmers or fishermen are frequent victims. Much the same applies to squamous cell carcinoma: but here the connection with ultraviolet light is even clearer. Generally, both of these cancers tend to strike older people; they are also quite likely to recover after treatment. By the age of 75, two out of three Australians will have been treated for skin cancer.

Cutaneous malignant melanoma is different. It kills. Forty per cent of sufferers die within five years. In three types, the frequency of its attack depends on the latitude. Once again, the further from the Equator, the less the risk. The connection is very precise. Even a shift of 2 degrees, or about 140 miles north or south in, say, the mainland of North America and the melanoma death rates change by 10 per cent. (There is a fourth type of skin cancer which occurs on the palms and the soles of the feet, which are not normally exposed to radiation, and this is the only type of melanoma found in people who have never moved south of the Arctic Circle.)

Malignant melanoma is on the increase. It is rising in all Caucasian populations, and striking younger and younger people. It doesn't seem to have much to do with long periods of weathering or exposure to the sun; if anything, short periods of intense exposure seem to be more damaging. There are statistics that should, once they become more widely known, put an end to the great summer stampede from the north of Europe to the beaches of the Mediterranean and Aegean. Girls (youth seems to be an important factor), for instance, who wear bikinis or sunbathe naked are 13 times more likely to develop melanomas on their bodies than those who wear one piece swimsuits. Adolescents with fair hair, fair skin and lots of freckles are 37 times more likely to contract malignant skin cancer than those with dark skin — richer in the natural blocking agent melanin — dark hair and few freckles. White US males are 19 times more likely to develop it than black US males. The highest rate of malignant melanoma in the world is found in Queensland, Australia: the sunshine state in a country almost overwhelmingly settled by Europeans, and a country devoted to the sun. The incidence there is 10 times higher than in the UK; four times higher than in the USA. How much of this is due to a thinning of the ozone layer and a corresponding increase in ultraviolet is not easy to answer.

In the last few decades millions of people who are less at risk for geographical and social reasons (young office and factory workers, for instance, from Glasgow or Aberdeen) have taken up the habit of annually travelling more than 1600 kilometres south, looking for hot sun so they can fry themselves gently in oil, during the hottest months. It would be astonishing if the cancer rates were not soaring. But the ozone scientists are fairly sure of the connection. It is estimated that a 1 per cent reduction in the ozone layer will increase the amount of ultraviolet-B radiation (the one that we are worrying about) by 2 per cent. This is called the radiation amplification effect. There is also a biological amplification effect. This means that 2 per cent extra radiation could cause a 2, 3 or even 8 per cent rise — it all depends on which expert you listen to — in some types of skin cancer. And if CFC use was to grow at 2.5 per cent per year, then the ozone layer would be predicted to dwindle by 10 per cent by about 2050 AD. These are only predictions. What measurements can be made suggest that the ozone layer might be thinning at a faster rate, especially in the middle latitudes. And since 1983, CFC production has been growing at nearly 8 percent per year. Our grandchildren may yet have to cower from the sun. Skin cancer is only the most studied of the impacts of a shrinking ozone layer. But there are plenty of other reasons for fearing a rise in the ultraviolet.

For instance, it can suppress the human immune system: this indeed may be one of the factors in the development of some cancers. It may thus encourage the spread of infectious diseases such as herpes labialis and cutaneous Leishmaniasis; at the same time making vaccination programmes less effective. (It also kicks the dormant HIV virus — the forerunner of Aids — into life, but only in laboratory test tubes.)

All these effects are going to be worse in the tropics; the low latitudes are already exposed to more ultraviolet. The effect of a 10 per cent ozone decrease there could be dramatic. And the effects will, of course, be felt most harshly just where medical services are least able to cope. Blindness of the avoidable or repairable kind is already a scourge of the poorer nations. Ultraviolet rays increase the incidence of cataracts: the clouding of the eye lens which leads to blindness or impaired vision for millions already. In the USA alone, there are now 600,000 cataract operations every year. In the next 40 years, unless something is done about CFCs, there could be 10 million *extra* cataract cases to handle. The rays could also cause an increase in keratoconjunctivitis, or snow blindness, a painful affliction that, fortunately doesn't last long. They can affect the growth of plants. Most species — especially at the seedling stage — are sensitive to ultraviolet radiation: they don't grow so well. Some soyabeans exposed to the radiation yield smaller, less tasty crops. None of this confirms the fears that forests and agriculture will suffer as the ozone layer thins. The higher plants, like animals (and the term animals includes humans) have a skin which absorbs ultraviolet radiation. Plants, like animals, can thus suffer tumours but not all those exposed develop tumours and not all tumours are fatal. But it is hardly encouraging.

There are going to be other direct effects. The gases in the lower atmosphere are going to be exposed to, and affected by, the kind of radiation that would normally be stopped 20 or 30 kilometres up. This means more ozone around us; more photochemical hazards. There will be more atmospheric pollution over wider areas. The people who are now struggling to improve air quality will have to struggle harder. Plastics exposed to ultraviolet degrade faster, too. So do paints, coatings, rubber products, wood, paper and textiles. Once again, the tropics — and the poorer nations — will suffer most: whatever ultraviolet does to materials, it can do even more effectively in higher temperatures and longer sunshine hours.

But the real hazard to the globe, the damage that may feed back into the greenhouse effect and accelerate climatic calamity, may lie once again in polar waters. First, a brief visit to the tropics. Although coral reefs are among the richest habitats on the globe, they are few, and limited in area (and they are under onslaught from pollution as well). But even in the coral-fringed lagoons, the waters of the tropics are blissfully, invitingly clear. You can see all the way to the bottom. They are clear because there are no phytoplankton. This is partly because nutrients are scarce, but partly because of ultraviolet radiation. Plankton are the primary producers of the globe. The varieties that fix carbon dioxide have to live in the zone of the waters into which light can penetrate. Although some waters absorb some wavelengths better than others, ultraviolet can certainly get through. But plankton, unlike the higher plants, don't have any protective skin to screen them against damaging radiation. Therefore they are more likely to be found in turbid waters, waters with high concentrations of nutrients and waters least exposed to ultraviolet light. The continental shelves off the north and south polar regions are, as a result, the richest waters of all.

If there are 100 billion tons of carbon fixed every year by plants, then the phytoplankton take care of about 66 billion tons. This matters powerfully in two ways. One is that almost every form of life in the sea is built on plankton. About 30 per cent of the world's animal protein is in the form of fish from the sea, and those herring and cod, snapper and tuna, eels, shrimps, lobsters, squid, sea urchins and molluscs, together with whales, seals and sharks, and all the seabirds from skuas and boobies to penguins and puffins, depend ultimately for their survival on the phytoplankton.

Nobody knows for certain what an increase in ultraviolet will do to the phytoplankton. But it is unlikely to benefit them. In the first place, they tend to 'bloom' in spring and autumn, that is, when there is light but not when the light is at its peak. This of itself suggests they may be affected by radiation. There have been a number of laboratory tests. These are not always a good guide to what happens in the natural world. But they do show that even a relatively short burst of solar ultraviolet radiation drastically affects a plankton's ability to move, to orient itself, and to develop. This means that it has less chance of getting out of the way of dangerous light. Plankton have to move up and down in the water: when there is little or weak sunlight, they move to the surface; when it is bright they descend to find the level that suits them best. In full sunlight, for instance, the pigments that help them photosynthesize are bleached. Further, although the tiny creatures may not be destroyed by UV radiation, they are vulnerable at certain stages in their development: the proteins from which they are built can be irreversibly damaged.

There is another aspect. There is a very important group of organisms which live in the soil and can fix nitrogen from the air. Cyanobacteria can fix nitrogen dissolved in water. Most of the higher plants cannot fix nitrogen from air or water, so without cyanobacteria fixing the nitrogen, in for instance rice paddies, the world would be a poorer, harder pressed place. The cyanobacteria alone, world wide, is reckoned to fix 35 million tons of nitrogen each year. This is five million tons more than the chemical fertilizer industry can manage. The active agent in the nitrogen fixing process in these organisms is an enzyme called nitrogenase (it is supposed that the entire world supply of nitrogenase could fit into a plastic carrier bag). This vital enzyme is controlled by light. It is switched on by ordinary, visible light. It is switched off by ultraviolet-B radiation. The impact on agriculture, through this means of a thinning of the ozone, doesn't bear a great deal of thinking about.

There are further reasons for hoping that the ozone layer doesn't thin much more. Fish, shrimp and crab larvae are damaged by UV-B radiation. Most of the species we harvest for food have eggs or larvae near the sea's surface where dangerous radiation could damage them. Somebody once did the sums for anchovy larvae on the North American Pacific coastal shelf. These larvae are found in the top few feet of the sea between June and August, when radiation levels are at their highest anyway. If there was a 16 per cent reduction in ozone levels, then 50 per cent of all 2-day-old larvae would die. More than 80 per cent of 4-day-old would die. Larvae at 12 days old would be wiped out. But then the damage would have begun before that. The same reduction in the ozone shield would cut primary production of the phytoplankton by at least 5 per cent. This fall in the primary food supply would work right through the food chain. It could, on its own, reduce the global fishery fish yield by six million tons per year.

And of course, if there were fewer plankton to take up the carbon dioxide man is pumping into the atmosphere in ever-increasing quantities, then so much more would remain in the atmosphere to feed back into the greenhouse effect. But then the stripping of the ozone layer and the greenhouse effect, although they are two separate problems, are inextricably linked. Nitrous oxides and CFCs are involved in making both problems worse. A thinner ozone layer will mean a greater production of ozone in the lower atmosphere, where it will also contribute to global warming. Almost everything about the destruction of the ozone layer promises to make the greenhouse effect worse.

Conversely, however, one effect of the greenhouse will be a cooling of the stratosphere, which might slow down the rate at which chlorine scavenges ozone. Or it might make things worse. The impact on the climate of a warmer lower atmosphere and a cooler upper one is anybody's guess. The other important greenhouse gas, methane, is also building up in the upper atmosphere and could actually take chlorine out of the sky, by reacting with it to form hydrochloric acid. This might lessen the ozone destruction while adding to the burden of acid rain. On the other hand, at least one scientist has pointed out, possibly with his tongue in his cheek, that the sulphur dioxide which we deplore as the prime source of acid rain could just possibly damp down the greenhouse effect by encouraging cloud formation. Clouds have a high albedo: they reflect light back into space, and could thus help stop the world from warming too drastically. It would be ironic if we took severe action on acid rain, and thus helped make the greenhouse effect even worse.

But we are now firmly in the clouds ourselves. The interaction of the ozone, acid rain and greenhouse problems is so complex nobody is in a position to be sure of anything. The lesson to be drawn from all this is that the machinery of the ocean, atmosphere and biosphere is wonderfully complex, subtle and intermeshing. Any action within it, any expenditure of energy, any expulsion of material creates a ripple that may echo gently through the whole system. If the system is designed at all, it is designed to absorb a myriad tiny shocks and respond to them with a mixture of a myriad tiny cushioning effects and counter insults. If it is designed at all, it was designed on behalf of the sum of creation, not just one of its parts. And one of its parts, humanity, has spent the last 200 years chucking an array of spanners into the global works: the spanners are simply being returned. This section of the book has principally concerned itself with two of these spanners. One is the greenhouse effect, which is probably inevitable. The other is the continued destruction of the ozone layer: even if we ban all ozone destroying substances right now (and we haven't) the ozone layer will still go on dwindling for another 50 years or so.

But this was supposed to be a simple, two-aspirin chapter. And anyway, although human responsibility for the destruction of the environment is quite clear, and the need for concerted global action to soften the global calling-to-account is transparent to almost all governments, the future is not, ultimately, in our hands. The world has its own ways of generating cataclysm, very effectively, and without any help from us. We may have devised the means to wipe ourselves out, and to take half of creation with us, but philosophers who take a long view — one spanning a billion years or two — may be permitted a cheerful shrug. A glance at what nature can do suggests that all our arsenals are not equal to an earthquake, and all our interference with the biosphere's machinery is as nothing to the swift obliterations that may have happened in the past, and all our polluting vanities are as nothing to a swiftly arriving asteroid, or a climatic hiccup into an ice age.

BOOK III

Nature strikes back

CHAPTER 11

The case of the missing summer

On 21 August 1986, a volcanic lake called Nyos in the north west Cameroon in West Africa suddenly gave a huge heave, and tipped 200,000 tons of water down its valleys. Locked in the water were at least three million cubic metres of gas with a weight of about 6000 tons. This gas freed itself in the upheaval, and 1700 people died immediately, overcome not by the weight of water but by the gas. The accident threw the world's disaster services and its volcanic scientists into confusion: for months there was no consensus as to what might have killed the villagers in its shadow. There was, for instance, almost no chemical damage to the vegetation, although some of the young fig trees were marked, as though droplets of acidic water might have condensed on them. The victims all died without a struggle. Some survivors reported a smell of rotten eggs: usually evidence of sulphuretted hydrogen, poisonous in any but the tiniest concentrations, and often present in volcanic regions, but scientists who tested the waters of the lake could find no traces of either chlorine or sulphur compounds forming there.

About eight months later, both British and American research teams who rushed to the spot delivered their verdict. The smells of hydrogen sulphide had been an olfactory hallucination, not unknown in such cases. Huge quantities of gas from the earth beneath had been building up, in soluble form, in the crater lake, until the water was saturated. Then something happened. A small quake, a storm, some convection, or just one last injection from below, and then the gas started releasing itself, setting the water heaving and foaming. You can see the same effect when you shake a bottle of champagne and then uncork it, or pour lager beer out too fast, or squirt soda from a siphon. The droplets that marked the young fig leaves were evidence of weak carbonic acid. The gas in the lake was simply carbon dioxide, the gas we breathe out, the gas plants use to build themselves. The water in the lake was simply water with carbon dioxide in it, the same natural sparkling mineral water that French housewives concerned about *crise de foie* and upwardly mobile British professionals concerned to show their sophistication buy in supermarkets to place on the dinner table.

It was just one of those things which no one had ever imagined could happen. Nobody knows why such a thing could happen at Lake Nyos, but not, as far as is known, in any other crater lake in the world. But it did happen. The 1700 farmers and villagers of the north-west Cameroon were quite simply gassed to death by the bubbles from a mountain lake of fizzing Perrier water.

But then Nature has always had the first card as well as the last, and quite often in West Africa. The world's first fission reactor — usually attributed to Enrico Fermi, at Stagg Field, Chicago — was not a human invention. In Gabon, on the West African coast, there is evidence of a natural, uranium fuelled reactor that went critical two billion years ago.

The rules for making a nuclear reaction are very simple (although the rules for managing one safely and efficiently are not). You just get two smallish quantities of enriched uranium and slam them together. The trick is to find them in enriched forms. Most uranium is in the 238 or 'safe' form. But there is always a proportion of the much less stable 235 isotope form which splits easily. Nuclear operators usually have to 'enrich' the uranium mixture, that is, increase the proportion of 235 to get a reaction going. What happens in a reaction is that an unstable atom of U-235 absorbs a neutron flying from somewhere. A neutron is simply a neutral particle in the nucleus of an atom: they tend to come flying out of unstable atoms. The U-235 splits into two and two or three more neutrons come flying out. If one of those lands in another atom of U-235 then suddenly the quantity of neutrons flying about is doubled or trebled. You have a chain reaction. And you have heat. If the two lumps of uranium slammed together are enough to start a reaction — that is, if there is enough critical mass — then you have a chain reaction on your hands. You need to have a thing called a moderator around to speed up or slow the reaction down, because if you don't, or if you have slightly more than enough, you have an atomic bomb on your hands; at least, for as long as you have hands.

At Oklo in Gabon in 1972 an open-cast uranium mine supplied ore to a nuclear processing plant in France. A French scientist became puzzled by the fact that uranium-235 was present in the ore in smaller proportions than normal. Out of 700 tons of uranium, 200 kilograms of U-235 was missing. He began the detective work and found that the accidents of Nature had anticipated Enrico Fermi at least six times in the Oklo uranium lode long before anything we recognize as life, let alone humans, had become established. First, there was the accident of the watershed, which kept concentrating uranium washed out of igneous rocks into ore at one point in Oklo. Then somehow the same process concentrated the ore in such a shape that as neutrons escaped, at least one of them was absorbed by another U-235 atom, and the reaction kept going. This simply does not happen in normal uranium lodes or seams, and if it did no mining engineer would ever be able to get near them. To reproduce this situation artificially required enormous human ingenuity and a basis of deep theoretical understanding that began with Newton and reached its threshold only with Einstein and Rutherford and Fermi.

But it happened, accidentally, at Oklo. And the accidents continued. The same process that assembled the natural atomic pile at Oklo kept it up and running. The water acted as a moderator to 'slow' the neutrons; it also cleared away the 'poisons' that slow or stop nuclear reactions, such as lithium and boron. In the end, the Oklo reactor did what all reactors do, shut down, having consumed the fuel it needed to keep itself going. This seems to have happened six times. From the start of its 'commissioning' Oklo may have been responsible for the production of about 15,000 megawatt-years of energy, and ran quietly for several hundred thousand years, consuming at least six tons of U-235, in at least six reactor sites, before it shut down and decommissioned itself. And, of course, Oklo seemed somehow to have solved the problem of the long-term storage of nuclear wastes. When Ecclesiastes observed, several thousand years before the discovery of nuclear fission, that there was nothing new under the sun, he spoke with some point.

Those two episodes are examples of things that we didn't think could happen, until they did. But improbable things and violent, disrupting events which are utterly beyond our control happen every day: some of them, such as earthquakes and volcanoes, we take in our stride, thinking in the queerest way that because they have happened in the past, they aren't likely to happen in the future. Thus the slopes of volcanoes are colonized, and communities build up, and are wiped out disastrously in the next eruption, mud slide or lava flow. Nuclear reactors are built on earthquake fault lines. We now know that some earthquakes are inevitable, because we at last understand the machinery that drives them, and we know that yet another earthquake on the San Andreas fault in California is inevitable soon. But some of the reclaimed Bay area of San Francisco is actually built on the rubble from the 1906 earthquake. Sometimes, wilful, even perverse curiosity destroys us.

On 23 May 1960, at Hilo in the Hawaiian Islands, the populace was warned in time to avoid an incoming tsunami, a tidal wave. It isn't as if the warning were delivered to say, Pevensey Bay in Sussex or Nantucket. The people of the island had seen tidal waves only 14 years before. One had taken 159 lives and caused $25 million in property damage. Further, any oceanographer could have told them, and no doubt at some point had done so, that tidal waves, because of their peculiar genesis, are not like ordinary, wind-whipped waves only bigger. The tsunamis, which are powered by earthquakes, carry an enormous body of water and they can travel at 800, 900 or even 1000 kilometres per hour: faster than a jet plane, and although they travel low, and are scarcely felt out in the open ocean, when they strike the shallows and enter confined bays, the waves have been known to build up to 60 metres in height.

Even so, when they heard the warning some residents of Hilo *actually rushed down to the waterfront to watch the wave come in*. The only explanation for their behaviour was that, even though they knew that it could happen, they somehow found the notion of the sea rising up higher than a tall building and smiting them at several hundred killometres per hour so improbable as to be unbelievable.

Evolutionary biologists and geologists and palaeontologists are a little more used to the happening of improbable things, because they keep reading the evidence in the accretions of time. If the chances of a plant or an insect crossing, say an ocean, and surviving in any one year are 100 million to one against, then over 100 million years such a crossing is almost inevitable. Right now, we are just beginning to face the strong possibility that we are about to alter the globe's climate, that we might be about to wipe out between one quarter and one half of creation, that we may even be on the verge of beginning to destroy ourselves or at least our civilization. We are right to worry about these things: we have, ultimately, no lien on the future, but we do have to worry about the world our grandchildren may inhabit. Even so, it is as well to keep a perspective.

But we know, because we have been looking for the evidence, that Nature has, many times in the past, altered the world's climate, sometimes with great swiftness and, of course, without our help; that Nature — it is simpler to apply this general term than to keep referring to blind circumstance and stress forces and tectonic cycles and solar rhythms — has in the far distant past wiped out the greater part of creation several times. She has changed the climate in a very short span of time, perhaps as little ás 20 years, and she may even have wiped out vast assemblies of species in no time at all, literally in a bolt from the blue. She has given us a taste of this power, just a brief hint. Several times in recorded history, and before the astonished eyes of meteorologists and the ashen faces of farmers, Nature has done dramatic, hitherto unthinkable things: such as simply suddenly cancelling summer, practically everywhere over the inhabitable globe.

Everybody has heard of the historic eruptions of Krakatoa, Vesuvius and so on. Mount Tambora on the island of Sumbawa in Indonesia is less well known, although it made itself felt everywhere. In 1815, it erupted, suddenly and violently, darkening the sun at noon in Java 500 kilometres away. At the end of the eruption, Tambora was 1280 metres shorter and 100 cubic kilometres of dust and pumice had shot into the skies. It probably put more dust into the stratosphere than any other volcano in the last four centuries, and as we have seen, what gets into the stratosphere stays there for years. It also begins the process of diffusing and circling the entire globe. Dust in the upper skies reflects sunlight out into space. If there is enough of it, then it can make a serious difference to the amount reaching the surface of the earth below.

In June 1816 there were 15 centimetres of snow on the ground in New England, and killing frosts in July and August. The summer was the coldest on record for 200 years. Geneva and Lancashire recorded its coldest July ever. In Zurich, the following winter, people survived on sorrel, moss and cat flesh. There were food riots in France, and a famine in Bengal. The fearful disruption and widespread hunger may even have triggered the worst typhus epidemic in Europe's history and a world-wide epidemic of cholera. It began in Bengal and spread slowly through the world, along the pilgrimage and trade routes, to afflict the USA 16 years later. Mary Shelley was inspired in 1816 to write Frankenstein.

One climatic historian has also suggested that the optical effects of volcanic dust in the stratosphere may have set the painter J. M. W. Turner on his unique course, and that the cold summers and bleak winters may also have turned women's dress from daring to modest and 'Victorian'; and that of the first nine Christmases in the life of Charles Dickens, up to 1920, six were 'white'. Dickens wrote very little about Christian religion, and when he did so he wrote both clumsily and somewhat awkwardly but he remains the greatest celebrant of the 'spirit of Christmas' in the English language. It is curious to think that our by-now-indelible idea of what Christmas ought to be like — sparkling snow, chilblained hands, hot punch and roaring log fires — may owe something to the random, sudden, fiery decapitation of a mountain on an island in the East Indies.

But volcanoes may not be random things, blowing their tops every now and then, or smoking away quietly for periods, occasionally bubbling fresh fields of lava before resuming their geological sleep or shutting down forever. They have been raising the dust for aeons: they have helped make our atmosphere what it is and they may have moderated or altered our climate in more ways than one. Some scientists think there may be a periodicity which might explain the ice ages — both the ones that last 100 millennia, and the 'little' ones that last only decades. The argument is that periods of major warmth may coincide with periods of low volcanic activity, and the instrument that sets the global thermostat may not be dust (although fine dust, sometimes as fine as a ten thousandth of a millimetre is everywhere, affecting the light) but fine droplets of sulphuric acid creating cloud cover in the stratosphere. One scientists who has been looking in the Greenland ice cores that date back, neatly, like the rings of a tree, has found natural, and probably volcanic, 'acid rain' signals that match the decades when weather is colder. The warmer years have correspondingly lower acid traces. But there must be more than one cause: another has been examining the match of dust, carbon dioxide and temperature. The more dust, the more ice, the less carbon dioxide. If there is 20 per cent less CO_2, and five times as much dust, then the temperature drops by $6°C$.

There is another, recent theory about volcanoes. This is that they may be the unheeding arbiters of creation: evolution's whistleblowers. The theory is that deep in the bowels of the earth, near the lower mantle, a layer of the globe's structure is unstable. Every 20 or 30 million years huge bubbles of molten rock rise slowly from deep within the earth to the crust, taking millions of years to break through, but doing so more or less coincidentally, making a violent series of 'super eruptions' which would alter the global environment so drastically as to change sea levels, darken skies and reshape landscapes, and make survival impossible for whole communities of species.

It is a theory pleasing to geologists, who once believed in an explanation of the globe called 'catastrophism' — that is, you could blame fossils of extinct reptiles on Noah's Flood and so on — but who for a century or more have been committed to a principle called uniformitarianism. That is, what is happening now has always happened, and what has always happened is enough to explain mountains, oceans, atmosphere, rivers, lakes, forests and by extension, animals and plants. There has always been a problem with the second thesis because it doesn't explain why, for millions of years, the globe should be dominated by families such as the dinosaurs, the huge reptiles, and why, suddenly, the geological record of tyrannosaur and apatosaur and icthyosaur and pterosaur, etched forever in the rocks and clays and sands and silts of the past, should, at a fine line everywhere in the world known as the Cretaceous-Tertiary boundary, and representing a sudden change 60 million years ago, come to a dead stop.

The 'bubbling mantle' propositon has a certain charm. But geologists and astrophysicists had perhaps got a little bored with the arguments for gradualism. Besides which, there have been a number of sudden, or at least apparently sudden global extinctions of species. In one, the Permian, 95 per cent of creation was wiped out. There have also been a number of attempts to fit these into some kind of pattern in time. If there is a pattern — and not everybody is so sure there is a pattern — it might have a period of very roughly 30 million years. There was, so to speak, a shadowy problem in search of a brightly-lit solution. About 10 years ago scientists revived, in an argument that turned quite swiftly into a jovial furore, the notion of simple catastrophism. The argument has been a source of anger to geologists and palaeontologists, and simple fun for the public, ever since. It is simply this: the end of the dinosaurs, and perhaps the creatures of other, earlier eras, may have come even more suddenly than any lurching of volcanic lava. It may have arrived, literally, as a bolt from the blue. The 'Thing', swift, obliterating, and world-altering, may have come from outer space. And it, too, or visitors like it, may arrive every 26 million years or so.

CHAPTER 12

A bolt from the blue

The geological record is just time's debris: in the ideal geological formation, exposed in, say, a canyon or a railway cutting, the soil at the top is the most recent and the bedrock at the bottom is the oldest. Geology is the art of reading backwards in time. Leap back a few billion years and there are the first rocks of the hot, barren surface of the Earth, and then on top of them is the evidence of sand and gravel and silt from weathering by water, swept down from higher ground, and then on top of that may be some limestone, suggesting that at one point the piece of earth you are standing on was under sea, and then next layer could be sandstone, suggesting that the sea retreated and left a beach behind, and the next bit might be silt, and you would know that a river estuary washed the spot for millennia. There might be basalt: to show that at some point a slow volcano lethargically belched its molten rock sideways over the same ground. After that there might be sandstone with ripples in it, like those made by wind forming dunes on a desert, and there might after that be more muddy silt, with coal in it: then you would know that once it had been a forested swamp.

In the ideal canyon the river has excavated history: it has cut through both soil and time's traces, and left the patterns on either side of the canyon for you to read. If such a canyon existed all the way from now to the beginning of the earth's first cooling, then about half way up, if you knew what to look for, you might read the first inscriptions of life in a silty shale, but it would be difficult. The bones of sea creatures — ammonites and shark's teeth so on — are more easily spotted further up; some famous fossil beds are extraordinarily rich in the relics of creatures great and small: whole icthyosaurs in one cliff, a Tyrannosaurus rex or an occasional archaeopteryx with feathers, sometimes spiders have been preserved for many millions of years, quite often ancient excrement, occasionally pollens.

It is possible, patiently, to chart the progress of evolution in a series of vertical steps. The very word evolution suggests gradual change towards greater complexity, as if with a sweep of a knowing eye up the side of the ideal canyon, you could trace life's development from simple, sluggish creatures through a series of clumsy approximations to fitness until we reach the present, where myriad forms are perfectly adapted to an equal number of niches with homo sapiens emerging after a series of false starts, into a world that might have been made for humans. That in fact, was how the late Victorians and their inheritors in this century tended to draw the picture of evolution from the fossil record.

In fact, it is much more complicated than that. In the first place, the geological picture is usually no more finished than a child's drawing, and like a child's drawing, it may have been folded, torn or scrumpled up more than once. In the first place, the sediments may be missing from the record in a number of places: swept away by flood, scraped by glaciers, or blown away in the hot desert winds. They may be altered dramatically by the pressure of the rocks above, or by the heat of vulcanism: shales and sandstones may be metamorphosed into schists and micas; limestones into marble. Intrusions — great, slow upwellings of molten rock from the earth's mantle — far below the weathered surface may split the strata vertically and form granite outcrops which muddle the picture. Earthquakes can shatter the pattern, and move two layers away from each other to create another kind of puzzle. Mountain formation can squeeze and buckle the layered sediments or even turn them upside down.

The fossil record disappears when rocks metamorphose, but then the fossil record was always haphazard. The soil is as much a part of the great universal co-operative friendly society as the air: most of life turns literally to dust and ashes, returning its constituent minerals, water, carbon and gases to be used again — and only occasionally, in swampy lakes and peat bogs, or sudden sterile deserts, or occasionally in the glaciers of the ice-caps, is a hint of the life of aeons past arbitrarily preserved, and bones converted by slow mineral accident to enduring fossil imprints.

Even so, after 200 years of patient puzzling, geologists have been able to build up a picture of the patterns of time, and they have found that although evolution happens — that is, adaptations get cleverer — the change is not continuous. Whole groups of creatures disappear. Whole new communities appear. The world's climate seems to change, quite suddenly, and the very character of the rocks of one era is markedly different from another. The epochs of time are distinctive enough to be given names, often devised from the geological regions over which the Victorians walked, with hammer and notebook: the Precambrian, the Jurassic, the Permian and Devonian, the Carboniferous in which we quarry for coal, and so on.

The end of each epoch is often seemingly sudden. This is quite frequently simply one of time's ruses. It may be that a strata which was evidence of a gradual shift from one era to another was eroded by the conditions in the following ages, making the break seem sharp. It may be that the evidence of change has simply been compressed, so to speak, by the burden of aeons above it, and the apparent jump actually took 10,000, or 100,000 or a million years to come about. It may be that an era seemed to come to an end in one part of the globe but continued for a few million years in another, and geologists are seeing a time-line, a punctuation, so to speak, which simply isn't there.

But one of these ages, the Cretaceous, which takes its name from the chalk laid down most spectacularly in the white cliffs of south eastern England, came to an end everywhere in the world at the same time, 65 million years ago, and was followed so sharply by the succeeding tertiary age, that the discontinuity can be traced almost everywhere that the rocks survive. Geologists call it the Cretaceous-Tertiary boundary, or the K-T boundary for short. The discontinuity is not simply one of the strata: one of the rocks that could only be laid down in one set of conditions giving way everywhere to rocks that indicate another set of conditions. These indications are that below that line everywhere is evidence of a warm, wet, hospitable world inhabited by dinosaurs. Above it there are none. It is as though creation had actually everywhere, at the same time, drawn an abrupt line, and said 'That's enough'. The line really is abrupt. Geologists and palaeontologists, who study a global record, an earth's archive that stretches back three or four billion years, normally regard 10,000 years as no time at all, so to speak. But the K-T boundary is different. Two of the many scientists who have been examining the evidence of what happened, once dubbed it 'The worst weekend in the history of the world.'

It would be proper at this point to say that not everybody accepts what follows. The demise of the dinosaurs has been a source of entertaining argument for more than 150 years. The first scientists called themselves 'natural philosophers' and they were quite often clergymen searching for direct evidence of God's hand in time's record: they saw the fossil skulls and teeth and claws of the vast, vanished reptiles as 'sports of Nature', or as God's way of testing their faith in the Biblical chronology, or the species that were left behind by Noah's Ark, which then perished forever in the Flood.

Long before Darwin's Theory of Evolution, these arguments had begun to falter, but after Darwin, the argument changed. Dinosaurs had become too 'successful': that is, too big, and too cumbrous, and with too little brain, and were overtaken and out-manoeuvred in life's race by nimbler, quicker-witted, more versatile mammals who in the end gave way to humans. Sometimes the arguments approached the risible. A leading palaeonotologist claims to have counted more than 100 theories for their end, each less convincing than the last. One, for instance, was that the vegetation changed and the dinosaurs, confronted by unaccustomed browsing, all died of constipation. Another, related argument was that the dinosaurs produced so much manure that all the vegetation changed and they starved to death for want of accustomed food. A supernova — a massive stellar explosion — might have showered the world with lethal radiation. The global temperatures might have risen too high. They might have fallen too low. The salinity of the sea may have fallen. Or risen. The oxygen content of the atmosphere might have changed. Many palaeonotologists sidestepped the problem altogether, arguing that it was probably not ultimately solvable. Since the geologists were getting nowhere, the physicists and astronomers stepped in, and in January 1980, announced that the dinosaurs and many other creatures had perished in one vast cosmic catastrophe.

The evidence for such a catastrophe is there, in the K-T

boundary. In the Appenine Mountains of Italy, between the white limestone of the late Cretaceous and the greyish pink limestone of the early Tertiary is a thin bed of clay laid down 65 million years ago. It was examined by an American geologist called Walter Alvarez, and his father, a physicist, the late Luis Alvarez, and two chemists. They found that this thin clay bed was 30 times richer in a rare and heavy element called iridium than other clays.

Iridium isn't found very often in the rocks of the Earth. It happens, however, to be abundant in meteorites, those sudden, unpredictable visitors from out there, out in the solar system or even perhaps beyond it, and meteorites — we see them suddenly, on clear nights everywhere, and call them shooting stars — have always peppered the earth, and may account for most of the traces of iridium we find almost everywhere. But here was a massive iridium overdose, and all of it concentrated in one thin strata, not just in Italy, but in the same strata in Denmark and New Zealand and in geological cores drilled from the ocean bottom. There were other signs. In the same strata, too, there were signs that the quartz that occurs in rocks had been 'shocked': subjected to enormous, sudden stress. There was also evidence (found later) of unusual quantities of soot in some regions, and of massive tidal waves in others, and everywhere an indication of sudden climate change. To find all this in one strata, and that strata a line between one world and another — a border at which not only the dinosaurs but 90 per cent of the zooplankton and many of the terrestrial plants disappear — made the conclusion irresistible. The planet must have been hit by an asteroid or a comet.

Whatever it was, it must have been a fair size. To explain that thickness of iridium, it must have had a mass of 500 billion tons. If it was a rocky asteroid, it must have been at least 6 kilometres across, and if it was a comet, which is mostly ice, it must have been about 28 kilometres across. Either would have arrived with a thump. The asteroid might have been travelling at 20 kilometres a second when it hit the earth. If it was a comet, the speed would have been 65 kilometres a second.

Such an event is not at all unthinkable. Small meteorites hit the Earth all the time. Most of them incinerate in the atmosphere. The cores of the larger ones survived to pulverize craters on the earth; many of them have been eroded or obliterated by glaciation, but some have survived. Others have landed in the deep Antarctic ice: they are regularly retrieved and studied because they are less likely to be contaminated with earth material. Some come from outside the solar system, a few are believed to have come from Mars: debris, perhaps from an even larger impact on the Red Planet. There is a belt of asteroids between Mars and Jupiter and there is a class of asteroids called Apollo which regularly cross the Earth's orbit. The chances of them hitting the Earth are very small, but over long periods of time even small chances come up trumps. About four of them are thought to hit the earth every million years or so. Four, that is, which are a kilometre across, and weighing about 100 million tons, and arriving at about 25 kilometres a second. We know what happens when they hit. They slam into the atmosphere and at that speed hitting even thin air is like hitting a brick wall. They generate colossal heat and the outer layers of the heavenly visitor start to vaporize. When the fireball hits the earth it generates an explosion roughly equal to the atomic bomb dropped on Nagasaki, multiplied five million times. The crater from such an eruption is about 10 kilometres across and the blast is enough to shoot many thousands of tons of finely smashed rock into the stratosphere, darkening the skies and lowering the temperature.

But the K-T asteroid (if there was such a thing) was of course 5,000 times more massive. The crater would be about 150 kilometres in diameter. There simply isn't a crater on earth that could match such a visitation. But then the earth is mostly water. It is more likely to have hit the oceans. If it did so, it must have whipped up tidal waves 8 kilometres high. The further from the impact, the lower the waves as their energy dissipates, but in Texas, there is evidence, right at the K-T boundary, of a sandstone bed which could only have been dumped, very suddenly, by a tidal wave or tsunami 50 or 100 metres high. No ordinary submarine volcano or earthquake could have raised a wave that high, that far from the presumed impact. The seas, too, everywhere around the impact must have boiled and flashed to steam in the same instant. Any creatures within hundreds of miles of the impact must have been obliterated in the shock, the flash or the waves of violence that followed in seconds. Life a continent away must have felt the impact, and trembled. What followed, according to some enthusiastic modelling by theorists, was a complexity of disaster, an orgy of overkill so energetic that the wonder is not that the dinosaurs died out, but that anything at all survived.

The multiple apparatus of worldwide death would have begun to be assembled in the fleeting second before the impact. An arrival of a body that massive and at that speed would have compressed the air and heated it to about 2000°C. This is enough to set nitrogen aflame. Nitrogen burning with oxygen would have produced several billion tons of nitrogen oxides which would start to convert with water vapour into nitrous and nitric acids, clouding the globe in hot poisonous rain. The acids would have leached toxic metals from the rocks and poisoned the fresh water with toxins of lead, arsenic, mercury and cadmium. But there is more. Having generated a welter of heat that would make full-scale global thermonuclear war seem trivial, the world would have been sentenced to what we now call a nuclear winter.

The amount of dust thrown into the stratosphere would have converted a warm earth — so warm, then, that there were forests at latitude 85 degrees, right into what are now polar regions and the seas were up to 200 metres higher than they are now — into a dark and freezing one. There is more. Such a dramatic, sudden darkening, such a headlong fall in temperature would have 'freeze-dried' the forests of the globe. (The heat of the impact would have baked many of them dry. The winds from the same impact would have flattened forests 1000 kilometres away.) The first lightning strikes would have started fires which would have raged everywhere. Geologists studying the iridium layer in both New Zealand and Europe have found between 100 and 10,000 times as much soot as would normally be expected. They interpret it as evidence of a single global forest fire. Up to 80 per cent of the earth's timber and foliage could have been destroyed. The soot itself — 70 billion tons of it may have been smoking into the atmosphere — would have made the cold and the dark, colder and darker.

Imagine it! The tropics converted to tundra by a curtain drawn over the heavens and the only light the blazing of the trees, and the only thing to damp them a rain of nitric acid. Of course the destruction of forests on that scale could only have had one consequence: a sudden dramatic rise in the carbon dioxide component of the atmosphere. An impact on that scale may also have triggered a paroxysm of volcanic eruptions everywhere on the globe. Thus the dinosaurs were blinded, smashed, pulverized, swamped by tidal waves, burned in forest fires. They froze, they starved, they were poisoned by toxic metals, scarred by acids and showered by volcanic ash, and waiting for the survivors when they came out of the long night of the asteroid — it was at least three months — was a global warming.

If the asteroid had hit shallow seas rich in carbonates then the carbon dioxide content could have gone up as much as tenfold, with a consequent temperature increase of as much as 10°C, for 10,000 years. Notice that in one split-second, unpredictable, random event, there have emerged all the themes of these chapters: acid rain, smog, nuclear winter and the greenhouse effect. The impact can't have done much for the ozone layer either: whatever finally crept out into the sun must have walked into a bath of extra ultraviolet. And perhaps some 70 per cent of the species inhabiting the globe 65 million years ago were wiped out.

There are a number of questions still unsolved. One is, why should even 30 per cent survive at all? And where is the crater? As in all such hypotheses, the revisions begin at about the same time as the confirmation. There is a buried crater 35 kilometres across, in North America. It dates back 66 million years, give or take a million (it has to be said that the precise date of the K-T boundary is not clear). It is the kind of crater a 10 kilometre object might make if it came in at a low angle and hit the earth a glancing blow. The impact would still have been catastrophic, but the destruction of the atmosphere would have been even worse.

The pattern of extinctions isn't crystal clear either. The dinosaurs exist nowhere above the time-line, but they were on the way out even before any impact. The mammals were on the way in: the old Victorian thesis that a larger brain to bodyweight ratio, warm blood and a healthy interest in rearing the young is a better prescription for survival than a small brain, cold blood and eggs left to hatch themselves has its supporters still. The percentage of species that disappeared at the time-line is also in doubt, simply because the fossil record isn't good enough. Some of the foraminifera — the tiny zooplankton of the oceans — became extinct well before the impact, some survived for a while afterwards. There may also have been more than one impact: one scientist has proposed at least four hurtling asteroids over a couple of million years or so to account for what seem to be 'steps' in the pattern of extinctions.

There were other dramatic changes taking place at the time of the impact anyway. The Atlantic Ocean was being born. Once Brazil and West Africa fitted neatly into each other (the match of shapes has always been irresistible and even before continental drift and seafloor spreading became intellectually respectable and now-proven explanations, geologists had identified glacier tracks in West Africa that stopped at the shoreline and then, 9600 kilometres away, continued from the Brazilian coast inland, leaving a deposit of rocks not normally found in that part of Brazil but certainly found in West Africa at the glacier's source). The birth of the Atlantic itself must have had a dramatic effect on ocean circulation patterns and volcanic activity, and as we have seen, one good volcano can darken the skies. The submarine lava flows which accompanied the opening of the oceans, together with the fall of ash from the skies and the huge ejections of sulphur from the craters might have altered the alkalinity of the oceans, accounting for massive destruction of the smallest creatures in the sea. None of these arguments mean that an asteroid didn't come crashing into the earth with dramatic effects: they provide a pattern of possible species collapse which was already taking place when something fell from the skies and precipitated the end of an era which was already rushing to its close.

But then in the last 600 million years, there have been at least five major paroxysms of extinction. At the end of the Permian, some 250 million years ago, about 95 per cent of all known living things came to a dead end. Since then, there have been mass extinctions every 26 million years or so (although the fossil record is so capricious that even the phrases 'mass extinction' and '26 million years' are matters of sometimes furious argument. In the first place none of the extinctions is as apparently sudden as the one at the K-T boundary. There are a number of possible explanations. One is that sea level changes are cyclic, and since for most of the earth's history, the sea has provided most of the habitat and certainly most of the fossils, changes in sea level could have a dramatic effect. The second is that changes in temperature are more important than in sea level, although they can't be separated easily, since sea levels fall because the temperature does, and more and more water is locked in glaciers on land. There is also the philosophical argument that biological systems are not ultimately stable: like business corporations or football teams or political empires, they begin by struggling to keep alive, then they come on strong, they make their mark, they dominate the game, they seem invincible and then, at first slowly, and then at gathering speed, they collapse.

But the catastrophists think, like the people who read the popular papers, that life's fate might be in the stars. There are several reasons why periodic extinctions might be due to astronomical reasons. Several astronomers are very fond of the idea that the Sun has an invisible companion star (they call it Nemesis) which disturbs the cloud beyond the solar system — astronomers call it the Oort cloud — in which the comets are formed. If there is a companion star, it must have a period of orbit in which it repeats the steps of a dance with the sun. It might repeat them every 30 million years or so, sending a volley of comets in the direction of the solar system. Or the solar system itself, which is rotating with the galaxy of which the sun is a minor star, somewhere towards the edge, may disturb the cometary cloud periodically. And then there is planet X, the one beyond Pluto, the one that has been proposed but never discovered, which might follow an eccentric path and clout into a belt of comets somewhere beyond Neptune every 28 million years or so.

One team of scientists deeply involved in the great 26 million year extinction show estimated that every 300 to 500 million years about a billion comets bigger than 3 kilometres across are pushed into orbits that cross the Earth's own circle round the sun. Out of that billion, 20 might hit the earth. There would also be minor showers every 30 to 50 million years: minor because there would only be a million comets crossing the earth's orbit. Of these, about two might slam into the earth each time. They also argue that there are 'peaks' in the pattern of formation of craters and these peaks roughly coincide with spasms of extinction over the past few hundred million years. It is an attractive thought. Comets have always been held by the superstitious to herald disaster, and behold, scientists suddenly say they do.

Unfortunately, we cannot call upon the Oort cloud as an alibi for the latest mass extinction, the one going on right now; the one in which, quite unmistakably, and at possibly even greater speed than may have occurred 65 million years ago, up to half of the species of the globe are being wiped out. That is being done by a species which has only been around in an organized form for 10,000 years, and which started systematically altering the face of the globe only in the last 2000, and which in the last 200 years has begun a process of exploitation that might be obliterating, right now, 100 species a day. We are helping this process along by inventing for ourselves some of the mechanisms that Nature may have used in a chance event every few million years: acid rain, ozone destruction, the greenhouse effect, and so on. But most of all we are doing it because we need the land, and the produce of the sea and if we need it, we take it, and our fellow species stand no more chance of survival than if they were hit by an asteroid at 25 kilometres a second. The irony is that we were not even meant to be here for long. We are in a period known as an interglacial. That means a warm period between two cold periods, or ice ages. This is another occasion for some perspective. The human race is quite old, not as old as most of the mammals, but in various forms, quite old. But the human as an ordered, civilized animal, with the skill to work metals, for instance, and plan crops, has only been in evidence since the end of the last ice age. That is, for 10,000 years. Ice ages tend to last for 100,000 years. There has been a pattern of them over the last few million years. The spell between them usually lasts for 10,000 years. We are thus almost due for another ice age, or even perhaps overdue. But we may even have disturbed the natural cyclic rhythm of the ages. We may have seen the last of the great glaciers.

CHAPTER 13

Under the ice sheets

Nothing is forever, nothing is constant. The Earth rotates round the sun and we call that complete encircling a year: the Earth spins on its own axis and we call one complete spin a day, but neither of these things is forever, or, while they last, even constant. Just as castles eventually crumble, and young human beings grow old, so will the sun fade and bloat and die, and so will the system of encircling planets (we call it the solar system) come to an end. We can compute the calendar for a century ahead, and we can calculate the tides — those manifestations of the moon's tow upon the oceans — for a century in either direction and be right to the minute, but over a longer period we will start to be wrong. The daily rotation of the Earth doesn't quite divide neatly into the Earth's annual rotation of the sun, which is why every four years we have a leap year to allow the calendar to catch us up; but that is slightly too much of a catching up, and every 400 years or so we skip a leap year to bring us back into line again.

We manage better than our ancestors, who were distressed (at various points in history, because the change was made at different times for different groups) by the sudden loss of 11 days: the Julian calendar named after the Roman Emperor Julius Caesar was eventually replaced everywhere by the Gregorian, named after Pope Gregory XIII, then leader of all Christendom. The confusion was simply because although Gregory changed the calculation of calendar, he kept the months. Thus in 1582, at first only in a few countries, 4 October became turned overnight into 15 October. But the very name October — it is Latin for eighth month — is evidence that someone had already been fiddling with the calendar. And the very name month comes from the four-weekly phases of the moon which help to perpetuate and enrich the confusion. There are very nearly, but not quite, 13 lunar months in a year, but for historical reasons we have stuck with 12. (Why not? There are two ways of measuring the waxing and waning of the moon: from its own point of view and from ours, because both Earth and moon are moving, and one way adds up to a period less than 28 days, and the other more.)

The Christian version of the Julian Year began with the Feast of the Annunciation, just after the spring equinox — the point in the year when the lengths of day and night equal each other, and then the days begin to grow longer — late in March, but the Gregorian Year was deemed to begin on 1 January, shortly after the winter solstice, the longest night of the year (and the shortest day). Not everybody followed the change: curiously, the financial year almost parallels the Julian calendar, and for the taxman the new year begins in April.

The confusions go on: we measure angles and tell the time in a number system invented in ancient Babylon, so that in a metric world we count the seconds, minutes and hours in multiples of six and ten: while we could recognize midday (on a clear day, it is the point at which the sun is highest in the sky) for centuries we couldn't compute it to the minute or the second, because there were no clocks that could measure in minutes or seconds; nor could we compute midnight to the nearest hour, because we had neither clock nor sun. (The medieval monks, who worked very hard on time measurement because they marked the day into periods set aside for devotion, divided the hour into three puncts: if you got there somewhere around the appointed punct you were punctual.)

The arrival of the accurate clock (still called by navigators a chronometer, a time measurer) opened the final phase of systematic exploration of the world. If you could measure the sun at full height (midday) and then check it against a clock set in Greenwich, England, you could work out your longitude as well as your latitude. It was rough at first. We measure degrees of longitude and latitude in minutes and seconds still; and the first serious scientific navigators were pleased to be right to the minute. But that would not be enough to find a small island: they needed to be within seconds. But the second itself was a problem. A second was one sixtieth of one sixtieth of a twelfth of a day, but suppose you had to deal in split seconds: how do you measure a second, and fractions of it, everywhere in the world? A second is now based on the atomic clock: atoms resonate far more precisely than any mechanical clock can match. We have, however, made the muddle slightly worse. We have, by basing the second on the pulsations of a caesium atom, removed it from the year. So we have to occasionally reset the clocks to bring the count of time back in step with the rhythm of the years. It sounds like one of those problems we could solve if we really tried: somehow we could devise a clock and a calendar that would accurately measure the process of the days and the years without needing adjustment, something we could do if (to coin a phrase) we could only spare the time.

But we couldn't, of course, because the solar system is not constant, and nor are the earth's seasons, nor even the days. To take the last first, the spin of the earth on its own axis is slowing. The tug of the moon on the earth's oceans — we call it the tide — exerts a dragging force, and the days are, imperceptibly, growing longer. Back in the Devonian period, 370 million years ago, the day was 21.9 hours long. We know this because we know that corals grow every day: they grow in the sun and then shut down at night, and leave daily 'growth rings' in the same way that a tree does annually. Corals are affected by seasons, too, however, so there are also marks in ancient corals to indicate the years. The corals of the Devonian period show a count of 400 days per year.

There are other changes. We say the Earth goes 'round' the sun, as if in a circle, but it doesn't. It orbits in an ellipse. An ellipse is the line you might draw around an egg laid longways on a piece of paper. If you tried to work out the centre of an ellipse you would see that the centre would not be the same distance from all points of the circumference. In fact, the Earth's orbit is not nearly as eccentric as the shape of an egg: it is nearer to a 'stretched circle', that is, a circle that has been flattened a little. But the effect is the same. In an ellipse, the Earth is sometimes nearer the sun, sometimes further away from it. But the ellipse is not stable, either. It changes shape — it oscillates — with time. The times involved are very long: about 100,000 years. Sometimes it is almost circular, sometimes it is more elliptical.

Thus over periods of hundreds of thousands of years, the

earth sometimes gets more light and heat from the sun, sometimes less: not much, but nevertheless it makes a difference. We define winter and summer however, not by the Earth's progress round the ellipse but by the Earth's tilt. The daily spin of the earth is not perpendicular to the plane of its rotation round the sun: it is at an angle. For half the year, that is, for half of the Earth's tour round the sun, the north pole is slightly tipped towards the sun and the south pole away from it: it is then summer in the north, winter in the south, with 24-hour sunshine at one pole and 24-hour night at the other. For the other half of the year, the situation is reversed. So far, so simple.

But nothing is simple in this sort of explanation. Ellipses aren't like circles. This must also mean that for part of the year the Earth is near the sun, for part of it a bit further away. The further it is from the sun, the slower it travels (this is a matter of Newtonian dynamics). Right now the Earth is nearer the sun at the northern winter solstice than it is at the southern winter solstice. We talk, politically, about the favoured, rich north and the impoverished south. In fact, it's true in a global sense. The arrangement — the difference both in nearness to the sun and the speed at which the Earth travels around it — means that the seasons are uneven. The northern hemisphere has a warmer winter and the southern a cooler one than fairness should seem to dictate. Strictly speaking, each hemisphere receives the same sunlight over the course of a year, but any farmer or even gardener knows that a milder winter means a longer growing season, and a baking summer is not necessarily an advantage. The northern hemisphere also has a shorter winter: shorter by seven days, simply because the Earth is travelling faster at that point. But the seasons themselves are not fixed. In the end, the apparent unfairness is cancelled by time. The seasonal points — the solstice when the sun is lowest (or highest) and the equinox (when spring gives way to summer or autumn to winter) — themselves creep slowly round the Earth's orbit, continually being lapped by the Earth, every 20,000 years or so, to even the score. It is called the precession of the equinoxes.

So we have a daily cycle which is shifting imperceptibly. We have an annual cycle which is shifting every 20,000 years or more. We have a change in the shape of the orbit which takes around 100,000 years to complete. It's enough to make the head spin. But then there is the other little complication. Imagine a spinning top. And watch what it does when it spins. It spins on its point — its axis — but the axis keeps changing, it keeps describing a circle. The circling movement is much slower than the spin but it, too has a cyclic rhythm. So too does the Earth precess upon its own axis. If you were standing on a platform above the solar system, and you had 20,000 years to watch, you could watch the Earth perform the same oscillation in its spin: the north pole would appear to move. Now we define north by the fact that is the direction of Polaris, the pole star; this cannot have been true 10,000 years ago. North would have been somewhere else, not so far away, but somewhere else. At the same time, the Earth's tilt towards the sun shifts. At the moment, the angle of the Earth is about 23.5 degrees off the perpendicular. This too is not permanent. Over a 41,000 year period it shifts from about 21.8 degrees to 24.4 degrees. When the tilt is at its greatest, summers would be hotter, winters colder.

So we have a number of cycles, every 20,000 years, 100,000 years, a second 20,000 year cycle, and a 41,000 year cycle. And each one of them makes a difference to the amount of sunlight received by a particular part of the Earth at a point in the course of a year. One of them — the 100,000 year change in the shape of the Earth's orbit — must affect the total sunlight over the whole globe: not by much, by about 0.3 per cent. Any change that small sounds inconsequential, but a shift in the balance of say 0.3 per cent would hardly seem of no consequence to, say, a man wheeling a woman in a wheelbarrow along a tightrope over Niagara Falls. The analogy may not be far-fetched: all the evidence is that the fabric of the biosphere is quite finely balanced and we may be about to see what a tiny shift in the atmospheric balance can do to the world's climate.

But all this is to assume that the sun burns with a steady, constant, inexorable brightness. It doesn't. The sun itself (which is also hurtling through space, locked in a giant galactic cycle, taking the Earth and all the other known planets with it) is not constant. It pulsates, resonates, convulses. Its fierce flames, fed by a colossal thermonuclear reaction, have cycles of their own. There is one we know best: the sunspot cycle. Over a period of 11 years or so, solar flares erupt with enough force to set the alarm bells ringing in the magnetosphere: a kind of shell of force fields and electrically charged particles round the Earth, at the outer edges of the atmosphere. We have become used to some of these phenomena. We can see them, the sun's eruption's reflecting in occasional ghostly glows at the high latitudes of our own planet. We call them the aurora borealis and aurora australis; the northern and southern lights.

The same storm forces can hit us in other ways: they can block out radio transmissions; they can heat the Earth's upper atmosphere, causing it to expand and slowly drag satellites out of orbit and down to their obliteration. They can 'knock out' navigation satellites. During one magnetic storm, two aviation satellites simply forgot where they were: their electronic brains were stunned by the sun's sudden force. During another, the Toronto stock exchange had to close down, its computers shut down by an arbitrary thermonuclear hiccup 91 million miles away. There is more than a suspicion that sunspots have a link with the world's weather. Scientists who have been examining tree rings in northern America think they have identified a period of drought that comes up every 22 years or so. Anything that pops up in multiples of 11 sounds more like sunspots than moonshine. This is not surprising. The sunspot cycles must, however slightly, affect the strength of the light radiating upon the Earth, and these too may be part of a larger cycle, in which, over hundreds or thousands of years, the sun itself waxes and wanes a little. In fact, the changes in the total amount of sun hitting one particular spot on the Earth over the course of a year during any or all of these cycles might be very small. The change may simply be a matter of a small adjustment between the temperature of winter and summer.

But small changes can have big consequences. We know this, because we have ice ages, big ones (we give them capital letters because they enfold huge areas of the globe in the embrace of ice for 100,000 years) and small (when farmers in northern Norway and Greenland abandon their land, as the ice advances a few hundred kilometres further south) and these ice ages too have a rhythm, and this rhythm seems inescapably tied in some way to the cycles we have been describing.

The picture is complicated. It is, first of all, complicated by things that happen on Earth. Water, for instance. Water stores heat. A wet world is a warm one. The Antarctic is colder than the Arctic (and the Antarctic ice line extends much further north than the Arctic ice line runs south) simply because the Antarctic is an enormous continent, and the northern pole is covered by sea, even though over the course of the year both poles get the same amount of sunlight.

Continents are different. The seas store heat and release it gradually during cold periods; land heats in the midday sun and surrenders the same heat swiftly as the night falls. In the Sahara or in central Australia you can bake by day and shiver by night. Russia, central Europe, the middle west of America and the plains of Canada see vast seasonal swings of temperature. England, Scotland and Ireland, and even Iceland, washed by the sea currents, have much milder climates than their latitude dictates. There is more. There is the albedo effect. Actually, at the summer solstice, when the day lasts for 24 hours, each of the poles receives much more sunlight than the Equator does: 1.4 times as much, but 80 per cent of the radiation is reflected back into space from the mirror of the ice. Ice is the outcome of cold, but it is also cold's begetter.

Conversely, there are inland areas — the very hearts of continents, where the temperature difference between day and night, winter and summer, is quite small. It is the albedo effect again. A blanket of trees absorbs heat from the sun as efficiently as ice or desert rejects it. In eastern central Africa, on winter's nights, the Zambians and Kenyans reach for their blankets and pullovers; the forest dwellers of the Congo and the Amazon are warm night and day, winter and summer, shivering only when it rains, because water conducts heat away from their skins so efficiently.

Thus what happens to climate depends on where the land is, and where the sea is, and whether each is covered by ice, and whether the land that is safe from the ice, is naked sand or cosseting foliage. Since the conditions on the continents themselves vary according to their nearness to or distance from the Equator, and since for the last few billion years or so the continents themselves have been scudding over the surface of the Earth, floating like scum on the conveyor belts of the ocean floor, colliding and melding into super-continents and fracturing and reforming (Africa and South America, India, Antarctica and Australia were once one big continent called Gondwanaland: Scotland was once part of Labrador; the Himalayas and Tibet are but a crumpling that began when India slid into what is now Asia) and since volcanic activity itself, the darkening of the upper skies with soot and sulphur, can dramatically change conditions, can stifle the sun's light and cancel the summer, or turn a cold winter into a big freeze, the picture of what triggered any one particular ice age may be beyond resolution. But most climatologists have a pretty good idea of what it is that makes a warm world tip quite suddenly into a cold one. It is, of course, the amount of carbon dioxide in the atmosphere.

In the bubbles of air, trapped in the ice sheets of Greenland and Antarctica, over tens of thousands of years, and in the sediments of the seabed laid down over hundreds of thousands, scientists have been able to read the records of climates come and gone. Roughly, the ice ages of the past (about 10 in the last million years or so) correspond to a 100,000 year cycle: the cycle of the change in the Earth's orbit from elliptical to circular and back again. The other cycles may play a part. They are not in step, so they may work against the first to mitigate its effect, or they may occasionally combine to make a particular ice age worse than another, more severe, more sudden, but these things probably don't play as much of a role as the conditions on Earth. There may be more than one effect at work anyway. Most climatologists will frankly say they aren't too sure about the mechanisms that precipitate an ice age: why should they be? All they can do is sift through the evidence looking for clues to something that happened 110,000 years ago.

What seems to matter is the role of the oceans in the high latitudes: at the ice line. And, more particularly, the plankton. If there is a little more light, and if the other conditions are right, they may flourish, taking more carbon dioxide out of the air, and carrying it with them in their bodies to the deep ocean: if they take enough carbon dioxide, the atmosphere cools and the seas begin to freeze. As the seas freeze, ever more sunlight is bounced back into space and the cooling goes on. The effect amplifies itself: positive feedback again. What evidence there is (and it is very messy) suggests that it really may work like that. First the shape of the Earth's orbit changes, with a corresponding tiny change in the amount of sunlight reaching the planet. Then the amount of carbon dioxide changes. Then after that, the extent of the ice sheets change. It works, too, in reverse. All the evidence is that the reverse is quite sudden. The last ice age seems to have come to an end 10,700 years ago. In 20 years, the Atlantic ocean changed from cold and stormy to mild. Southern Greenland warmed 7°C in less than 20 years.

One should not take the possibility of an imminent ice age too calmly. During the last one, the sea levels fell by more than 100 metres. They fell because all that area of water (just imagine it! the first 100 metres of all the oceans; oceans which cover almost 70 per cent of the globe!) collected on the land and froze, expanding by 10 per cent or so, to make a series of great sheets of ice, in some cases 3800 metres thick, over Canada, the northern part of the USA, Russia and Northern Europe. In an ice age, the temperatures of large parts of the northern regions might never rise above freezing point even at midsummer. This would mean that the depth of ice would keep building. But a differential weight of ice could only mean movement: a movement southward. No fixed structure devised by man is likely to survive for long in such a movement. Most mountains bear the marks of massive gouging by glaciers. The seas, too must have frozen over large areas, but freezing seas on their own make no difference to sea levels: the ice displaces its own volume of water. With a fall of 100 metres, the Bering Strait between Siberia and North America dried up: we owe the colonization of the Americas to the last ice age. Humans — we imagine people of the time as hunter-gatherers, already skilled in the use of some tools, already aware of art, already clothed in skins and travelling in small bands — simply walked there.

Meanwhile, the evidence of the rhythms of the past, also suggest that in theory it may be about to happen again. If there is a climatic clock, counting out the globe's meed of long glaciations and short warm periods, then we should brace ourselves. The warmest part of this particular ice-free patch occurred about 7000 years ago. Until very recently, climatologists were more worried about the possibility of the return of an ice age than about a runaway greenhouse world. From 1940 to about 1970 the world appeared to be cooling. The possibility was that in five, 50, or 1500 years, the ice would return, as the astronomical clock that controls the tiny variations in the sun's benison tipped the balance towards the big freeze again. The possibility is still there, of course. Nobody knows enough about the cycles of climate to be sure.

As we saw at the start of this section, Nature has the joker, the wild card, and always keeps the deal. As we saw earlier, the evidence of the past is hard to read, but easy to interpret wrongly. As we saw at the start of this chapter, despite several thousand years of practice, we have still trouble telling the astronomical time. We are not even sure about how the Earth works, let alone the heavens. We have also, over the last 200 years, interfered with the springs of the mechanism: we have begun to increase the carbon dioxide in the atmosphere to a point where it could begin to parallel the climate of the Cretaceous, when Greenland and even some of Antarctica itself was forested, long before the present cycle of the ice ages. A fall of 4°C would bring the deep, everlasting ice back to London or Chicago, but we seem to be about to provoke a global warming of 3-4°C.

We seem to be poised (for once the cliche has a fitting feel to it) between the devil and the deep blue sea: a world of ice or a world of rapid warming. Or just possibly, the two effects may cancel each other out: proof that everything is for the best in the best of all possible worlds. But that would be counter to Murphy's Law, which assumes that whatever can go wrong, will: a useful rule of thumb in all human affairs. And in the end (this being the section devoted not to our role, but to Nature's) it may not depend on anything much that we can do. What happens may depend upon how the oceans work, and what role the plankton play. We have been here before. Once again, the tiniest creatures in the sea, the plankton, may decide the fate of the globe: may pump the cycle of change, and control the cover of ice and the height of the sea, may decide whether the oceans wash over New York or London, and convert Leeds or Chicago into swamp, or empty the basin of the oceans to bring down the ice sheet once again and grind all four cities into dust. Although, there again, humans may have it in their power to cancel the greenhouse effect and precipitate a long period of cold and dark, not unlike the calamity that overtook the dinosaurs, and at the same time grind their own cities to dust. All it would take would be one good all-out thermonuclear war.

CHAPTER 14

The message of Mars

In 1971 a team of planetary scientists settled down to watch the results from the first orbit of another planet: the encircling of Mars by the US satellite Mariner 9. They had hoped for at least three months of continuous observation: this was because the space engineers had guaranteed the satellite for three months but not, they thought, much longer. There were a number of reasons for a close look at Mars — one of them was to see if there were any Martians, or at least some form of Martian life, although most scientists had begun to discount that possibility long ago. Another was to see what could be learned about the nature of planetary atmosphere. The major gases on Earth are nitrogen, oxygen, water vapour and argon: carbon dioxide, along with neon, helium and krypton are very minor gases, but we know that the balance of the gas mixture on Earth has been altered (over the last two or three billion years) by life.

A study of Mars, and our other nearest neighbour Venus, might reveal something about the original nature of the atmosphere on Earth. The atmosphere of both Venus and Mars, for instance, is very largely carbon dioxide. On Venus there is also a small amount of nitrogen, as well as water vapour, sulphur dioxide and argon. On Mars the atmosphere contains some nitrogen and argon, as well as oxygen, carbon monoxide and neon. There are important differences. The atmospheric pressure on Venus is 90 times that on Earth, and the surface temperature is about 450°C hotter. On Mars the atmospheric pressure is 150 times *less* than that on Earth — quite simply the air is much thinner — and the average surface temperature 70°C lower which means that during the Martian polar winters, ice-caps form, although not of water but of carbon dioxide, the 'dry ice' beloved of rock video producers and horror moviemakers.

There are also similarities, although none of them is obvious to non-scientists. One is that if all the carbonate rocks on Earth — the chalks and limestones and other sedimentary rocks, including coal measures and oil bearing strata — were to surrender their carbon back into the atmosphere, then this planet, too, might have an atmosphere mostly of carbon dioxide. The conclusion is that living plants (which Earth has and Mars and Venus do not) removed the carbon dioxide from the atmosphere and released oxygen, and that oxygen breathing creatures consumed the plants and stored the carbon in their bones and shells, which gradually became geological formations. (The estimate, for those who enjoy numbers so large as to defy comprehension, is that 10,000 million million tons of organic carbon has been stored in the Earth's crust since photosynthesis began.) It is another reminder that even the inanimate rocks we hew for building stone or crush for cement are part of the building blocks of creation: the biosphere's rubble, so to speak, laid down over millions of years to pave new highways for the next generations of life.

All three planets, because they have greenhouse gases in their atmospheres, experience a greenhouse effect. Mars, with its thin atmosphere, is about 5°C warmer than it would be if it had none; Earth about 35°C warmer, and Venus about 500°C warmer: Venus, like Earth, also has clouds of concentrated sulphuric acid particles everywhere. On Earth, however, most of these are washed down by rainfall; on Venus the temperature is too high to permit water vapour ever to condense. Consequently Venus is covered with clouds of boiling sulphuric acid.

There are other similarities. Both Mars and Earth have orbits which change with time: it might be that Mars, like Earth, suffers periodic ice ages, although since the atmospheres are very different, the impacts would not be the same. Even so, while the atmosphere and surface temperature of Venus — that exemplar of the runaway greenhouse effect — made it finally clear that Venus could never be habitable, the similarities between Mars and Earth were enough to keep alive hopes of some form of Martian life: there might not be life on Mars now, but there may have been at some point in its past. Once the atmosphere on Mars may have been much denser; and there are scientists who believe that that atmosphere may have been stripped away by an asteroid, in a collision or a glancing blow of the sort that may have changed life on Earth at the end of the Cretaceous. There is, for instance, evidence of some kind of flow or drainage pattern on the Martian surface: there may have been running water, itself an indication of a once warmer, denser atmosphere.

Anyway, the similarities were enough for some scientists and space enthusiasts to suggest that Mars could, eventually be colonized: that it might be possible to 'seed' the Martian atmosphere with algae or plant life which would begin the long process of changing the soil structure and enriching the atmosphere with oxygen. But that is not the point of this catalogue of similarities. The thing that matters here is that both Mars and Earth suffer from dust storms.

But a Martian dust storm is not at all like one on Earth. In 1971, the Mariner scientists were horrified to find one raging on Mars as the satellite began its encircling of the Red Planet. It was a storm so intense it filled the entire atmosphere of the planet with blinding particles of dust. For the entire three months of the guaranteed life of the satellite, the Mariner scientists were confronted by a series of photographs of what Professor Carl Sagan called 'an entirely featureless disc'. In fact, the dust storm cleared, the satellite operated for more than a year and the planetary scientists were eventually able to examine the Red Planet from pole to pole.

But during those three months practically all that the scientists could examine were the data from an infra-red interferometric spectrometer: in other words, even if they couldn't see, they could read the temperatures of the Martian atmosphere from the outer edges, where it faded into space, right down to the ground surface. In theory, the temperature of a planet is highest near ground level and cools swiftly with altitude; atmospheres are tricky things, however, and although this is true in any area where man can breathe unaided, things get a bit more complicated after that. Thus simply because the Earth has an ozone layer, the stratosphere tends to warm with altitude. Neither Mars nor Venus has an ozone layer so they don't have the equivalent of a stratosphere either. All of which is why scientists expected the temperature of the Martian atmosphere to fall with altitude.

What they found was that the atmosphere of Mars was warm during a dust storm, but that the surface temperatures were extremely cold. As the dust settled, the surface became warmer and the atmosphere cooler. What was happening was that the sun's rays were blocked and absorbed by the dust, creating a period of abnormal cold on the surface below. Struck by this, Professor Sagan and his colleagues began calculating ways of applying these observations to the Earth's own atmosphere: they were at the time interested in the consequences of large volcanic eruptions on the Earth's climate. Indeed, it was at about that time that scientists began matching worldwide 'bad summers' to known volcanic eruptions (in the course of which they found evidence of volcanic eruptions hitherto unrecorded).

Within a few years scientists had started applying the same calculations to another kind of cataclysm, a global thermonuclear war. But first, some geopolitics. Up until the late 1970s, nuclear peace had been kept, however uneasily, by a proposition known by the acronym of MAD, Mutual Assured Destruction. That is, both Eastern and Western superpowers maintained huge arsenals of nuclear weaponry, with a total explosive yield of 15,000 million tons (or metagons) of tri-nitro-toluene, or TNT. For a measure, the bomb dropped on Hiroshima had a yield of 12,000 tons of TNT. The assumption on which the Cold War rested was that if one side started a nuclear war, the other would have enough weapons to retaliate massively: there could be no 'winners' in a nuclear war.

It is in the nature of such propositions that because they suit political purposes they are often first eagerly adopted, then adapted, and then finally questioned. In the early 1980s dangerous questions were being asked:

- Was it true that nuclear war could not be contained to one 'theatre'?
- Would it be possible for one side to launch a sudden, decapitating strike, that is an attack that would cripple the other side's battle control stations and neutralize its missiles?
- Would it be possible to devise defensive missiles that would strike down incoming warheads?

- Would it be possible to mount a huge defensive umbrella in space?
- Although capital cities and known defence targets would certainly be destroyed, and although millions might die and many more millions suffer or even perish in the aftermath, would there still not be many millions of survivors, especially in a population trained in civil defence?
- Would a nuclear war not be in some sense survivable, even for the combatants?
- Would there not be many parts of the globe which would escape not only direct attack but also the consequent radiation poisoning?

These questions were dangerous not because they were invalid (they are perfectly valid) but because they reflected a subtle shift in attitudes among the politicians and the military: a shift from 'nuclear war is unthinkable and the other side knows that too: that's why we need all these nuclear weapons to deter them' to something much more dangerous, an attitude that had begun to permit the question 'Maybe they think they can take us out and stand the cost and survive? Maybe we should think about taking them out now and making the world safer in the long-term . . .'

The findings of Sagan and his colleagues — they were published in the US journal *Science* on 12 December 1983 under the title 'Nuclear Winter: Global Consequences of Multiple Nuclear Explosions' — took the debate onto a different plane. They proposed that a nuclear winter would be a greenhouse effect in reverse. Under the greenhouse effect the atmosphere acts as a window for the light coming in, and a blanket to slow the heat escaping. A nuclear winter would have the effect of smearing the window to stop the light arriving. The blanket would then not serve its function.

The argument starts with the fireball from a one megaton bomb exploded at an altitude of a kilometre: this is intense enough to start spontaneous fires over a region of 10 kilometres. The blast that follows is also enough to put most of them out a second or two later, but in a city, which lives by petrol and electric equipment, the fires would break out again almost immediately. Most of these fires would be likely to merge into one big blaze: a firestorm, which would create its

The nuclear winter (appeared in the *Guardian* 9 August 1984; reproduced by kind permission of Peter Clarke)

own wind and fan its own flames to a point where concrete and steel, too would ignite. This happened in Hamburg and Dresden during bombing raids in the Second World War. In a modest thermonuclear war, one in which only 5000 megatons of warheads were exploded, the fires would hurl more than 100 million tons of smoke into the atmosphere: it may be as much as three times that figure.

If 100 million tons of smoke were injected into the atmosphere and spread uniformly round the globe, the scientists argued, the intensity of the sunlight reaching the ground could be reduced by 95 per cent. This would happen in any case over the areas in the target zones: the farmers who burnt their stubble in the late summer could, if the weather conditions were right, subtly darken the skies over whole counties in the UK, and forest fires in Australia can dim whole states. But depending on wind and rain, smoke in the lower atmosphere clears in days or at least weeks. To affect the whole world, a large proportion of the smoke needs to ascend into the stratosphere. The experience of very large fires is that they create their own updraught: they can funnel their own ash and soot as high as 20 kilometres, that is, into the stratosphere. What goes into the stratosphere stays there for months or years: it is, as we have seen, a somewhat separate world. And a nuclear exchange is inevitably going to cause very large fires.

But many of the nuclear exchanges would involve direct hits: explosions at or below ground, as the combatant nations aim for each other's 'hardened' targets — missile silos, strategic command posts and industrial centres. When a one megaton bomb explodes on the Earth's surface it shifts between 100,000 and 600,000 tons of soil, and about 10,000 to 30,000 tons of this gets into the stratosphere. None of this is conjecture. Up until the world-wide ban on atmospheric testing two decades ago, the US, British, French and Soviet atomic and thermonuclear scientists had plenty of opportunity to observe what happened upon detonation. The consequences are a function of size. A small warhead of 100 kilotons — an explosive force equivalent to that of 100,000 tons of TNT — injects almost nothing into the stratosphere. But the higher the yield of the explosion the greater the proportion that reaches the upper layer. It is estimated that beyond the megaton yield, most of the debris will reach the stratosphere, simply because the intense energy of the blast will create an intense, urgent, lift skywards.

The Sagan team thought it quite likely that in a full-scale nuclear exchange in which each side aimed not for the cities but for each other's air bases, missile silos and submarine pens, there would be 4000 megatons of direct hits: this would lift 120 million tons of tiny soil particles into the stratosphere. It would, of course, be joined by 100 million tons or more of smoke from forest fires. This would be enough for a climatic catastrophe. Quite how swift or final or world-embracing the catastrophe would be depends on a number of things, including the season of the year at which the war was fought, and the air movements in the stratosphere, but probably quite soon, to an observer from another planet, using a version of an infra-red interferometric spectrometer, the Earth would look remarkably like Mars in a dust storm. The upper atmosphere would heat by between 30°C and 80°C as the upper atmosphere absorbed almost all the sunlight. The ground below would cool in the darkness. Over the combat zones it would be too dark to see, even at noon.

A temperature drop of a few degrees over the northern hemisphere in spring or summer would be enough to destroy harvests: the world learned this during the 'year without a summer' in 1815-16. After a 5000 megaton nuclear exchange, the temperature drop in the northern hemisphere would be about 40°C. That is, the temperature would fall to minus 25°C. It would not, of course, be the same everywhere: the oceans are a vast reservoir of heat, and would surrender it slowly, but inland continental temperatures would fall with astonishing speed. Without the heat of the sun to drive it, the movement of the air itself could cease. It could be months even before weak sunlight began to filter through but by that time almost everything would be dead. The disaster would extend to the Tropics and to the southern hemisphere quite rapidly: the sheer scale of the disruption would probably alter stratospheric circulation patterns and take calamity to the innocent — that is, the noncombatant — south within a matter of days.

There would, of course be other, very well-known consequences of a nuclear war: widespread immediate death and enormous suffering; almost utter disruption of water and food and power supplies and medical services; communications would close down because of a phenomenon known as the 'electromagnetic pulse' — that is, airborne explosions would provide a charge which would overload electronic instruments; the globe would be showered with radioactive particles which for years would provide a steady, toxic rain upon living things; the effects of 1000 or more blasts over industrial zones would be to release huge quantities of toxic chemicals now contained in secure factory stores into the environment. The very buildings themselves would, as they blazed, release carbon monoxide, cyanides, vinyl chloride, dioxins and furans, all poisonous. And of course, the fireballs would ignite the nitrogen in the air, creating oxides of nitrogen which would damage the ozone layer and increase by several hundredfold the amount of biologically dangerous ultraviolet light in the atmosphere.

The blast, fire and radiation alone could cause about 1.1 billion human deaths (equivalent to the population of China) and about 1.1 billion serious injuries. Under the nuclear winter hypothesis the sufferings of the remaining 60 per cent of the human population would be according to the season at which the war was fought. There would be widespread burning of forests during a nuclear war; there would also be widespread forest deaths in the months afterwards because of radiation fallout, as well as the intense cold, but that, too depends upon the season at which the war would be fought. A dramatic fall in temperatures in midwinter, when northern hemisphere plants are dormant, would be less damaging than one at spring or early summer, when all plants are vulnerable to sudden freezing. A pine tree accustomed to Alpine climates, for instance, can survive a temperature drop of $-50°C$ in the winter: but collapse at $-5°C$ in the summer. The pollen of rice and sorghum become sterile at $13°C$; corn and soyabeans are affected by a summer fall to $10°C$.

The combination of low temperatures and low light levels would shut down agriculture practically everywhere on earth for months. Hardy perennial and temperate plants would probably survive — or at least their seeds would — but the tropical plants might not. The animals that grazed upon either would be unlikely to survive. The phytoplankton upon which all marine life rests would suffer: nobody can even begin to guess by how much. Those plants and animals that did survive would, when the skies cleared, be subjected to massive overdoses of ultraviolet B radiation. Birds, too would die in enormous numbers. There would, however, be a class of life less likely to suffer. Insects and their parasites are affected less by radiation and climatic catastrophe than most forms of life and they would find in decaying vegetation and thawing corpses, the means of survival denied all other living things. Animal survivors who held on through the fireballs, the poison rain, the long cold night and the blistering radiation that followed, who held out against starvation, foul water and the moral despair that would accompany the shutdown of the world, would stagger out into the light to meet not hope but plague.

And all this from a nuclear exchange which used only one third of the available arsenals of the world. The only good news from the nuclear winter study was that such a cataclysmic blow to the environment would be unlikely to trigger another ice age, largely because of the enormous heat storage capacity of the oceans. The impact of the findings was not immediate (they were published, for instance in the *Guardian* newspaper in London by one of the participants, Professor Norman Myers, an environmental scientist who lives in Oxford, weeks before their formal disclosure in the respected columns of *Science* and were received in almost total silence). It took months before the findings began to filter through to the politicians, the military theorists and the rest of the scientific establishment; curiously, the conclusions were then subjected to almost instant attack.

In some cases the questionings were perfectly proper: the calculations were based on processes in the upper and lower atmospheres that nobody really understands, and on hypotheses about nuclear war which are unproven. In some cases the attacks were oddly petulant: while the findings were in strictest terms unprovable (except by running a full-scale thermonuclear war and observing it scientifically, obviously an impossibility) they were a gift to the swelling peace movements of Europe, the USSR and North America which adopted the conclusions uncritically. And as any observer of the recent Cold War will have noted, the official defence establishments of the West actually get on quite well with the defence establishments of the East; but both sides have tended to regard their own peace movements as dangerous enemies to be defeated politically or by prosecution.

Even so, the nuclear winter predictions have been a source of scientific argument since 1982: the debate is not finished but the consensus, accepted by some of the original researchers, is that they may have overstated the case. The notion of a nuclear winter has been adjusted to a nuclear autumn; although a nuclear autumn during a northern hemisphere spring would itself be a global calamity, accompanied by widespread famine and disease and a blighting of the vegetation, as well, of course as 1.1 billion deaths and 1.1 billion casualties. There were other reasons for the bleak shrugs that in some cases greeted the findings. One of them, prevalent in Europe, the likeliest theatre of combat, was that since nuclear war meant utter death and destruction on the spot, as well as the collapse of a 2000-year-old civilization and all the values invested within it, the prospect of slow death and destruction in the course of the following year wasn't of itself particularly frightening, not to a community of nations that cynically expected to be obliterated anyway before the two Superpowers tried to stop the nuclear exchanges.

But paradoxically, as the scientists themselves began to accept that the long term consequences of global thermonuclear war wouldn't be quite as appalling as they had calculated — for there are degrees even of utter catastrophe — other forces began to see the main point of the research. That is, that there would, after all, be no 'winners' in a nuclear exchange. The idea, uncheckable, subjected to endless wrangling at all its premises, eventually permeated to the upper reaches of power on both sides of the now partially razed Iron Curtain. In some cases, it was seized upon eagerly: most Cold War 'hawks' have at one time or other admitted to feeling deeply uneasy about the buildup of nuclear weapons, and the idea of a nuclear winter, even while it was being revised, entered the language of disarmament. It almost certainly contributed to the mood which permitted the agreement on the dismantling of cruise and other intermediate missiles in Europe, and the steady lessening of the tension between the Superpowers. It also contributed to an apparent slowing of the arms race, which may eventually yield incalculable economic savings on both sides. The weapons still exist, and there are still weapons laboratories planning and refining the next generation of superweapons but, for the first time in more than 40 years, the fear of war between the Superpowers has receded dramatically. And when fear departs, sweet reason returns. Thus the notion of the nuclear winter may, in a small way, have contributed to the whole range of new attitudes in the Soviet Union, which, too unwieldy to change swiftly itself, began signalling the possibility of dramatic change to its neighbour states. In the closing months of 1989, against all historical experience or expectation, the USSR suddenly permitted or perhaps even encouraged a series of revolutions in Eastern Europe, all but one of them more or less bloodless, which ended in the tearing down of the Berlin Wall and the stretching of friendly hands across the Iron Curtain. If all that is so — and if this new, happy state of affairs lasts — then it has turned out not to be such a bad pay-off for three months of staring expensively across a few million miles of space at the featureless surface of an invisible planet, shrouded in an undreamed-of storm of dust.

In the wasteland

CHAPTER 15

The great extinction

Up till now, this book has been dealing with what could happen: the cards that could be played from the shuffled deck of the future. I have tried not to make it grim reading, for two reasons. All the processes described so far are really only adjustments in the machinery of nature; adjustments made by us to be sure, but humans are animals, and part of nature, and our impact on the greenhouse effect or the ozone layer is, though greater and more complex, in the end like the impact of elephants upon the African bush, or wild sheep or rabbits on a grassland hillside: all creatures alter their own biosphere a little to their advantage, and sometimes to their detriment, and big and numerous creatures make greater and more apparently permanent alterations.

There is another reason: the future is always a little frightening, but when it arrives it is merely the present, and somehow we have adjusted to it, made the best of it, survived. Further, the future never arrives in quite the form predicted: not so bad in some ways (because we have prepared a little); rather worse in other, unexpected ways; and every now and then, for someone, or for some group, it also brings a bonus, a windfall. And anyway, we really ought to be more frightened by the present. So many things have gone seriously wrong already. The human population is soaring. Because of this, the areas of land under forest are shrinking, the areas of land classified by geographers as deserts are increasing, and the seas are becoming at the same time more polluted, and less rich in fish. Because of these four things alone the world faces famine, and not just wider, more frequent, more protracted and more sickening outbreaks of hunger, malnutrition, disease and starvation of the sort that has scarred the lives of millions in India, China and Africa many times already this century, but a deeper insecurity, one that puts at risk the food supplies of the more comfortable nations in the temperate zones. These four things are what politicians and economists call 'destabilizing': they increase the risk of political upheaval, civil bloodshed and open war, which in turn step up the dangers of famine.

Attendant upon war and famine of course, is plague. The words have about them an apocalyptic ring. Bloodshed, plague and famine were the burden of three of the four horses of the Book of Revelations, the New Testament vision of St John upon the island of Patmos, and indeed the words sit comfortably upon the page of a book, written in the closing years of a millennium, that speaks of the end of an era, that outlines a kind of calling-to-account for the human race. But they are still fanciful. Bloodshed, famine and plague are nothing new in human history, nor in Nature (that is, in the rest of the natural world). What is new, and what is a consequence of actions, taken, quite unheedingly, now, this century, unlike any other, is a kind of apocalypse nonetheless. It is the widespread, wholesale extinguishing (that is the blotting out or erasing) of quite possibly half of the species with which we share this planet. And not the least cruel aspect of the extinctions are that we don't even know what we are doing. It is as if we were committing genocide, systematically and calculatedly committing whole genera, that is, groups of species, to some kind of animal Auschwitz, without even being aware of who we were destroying, or at what rate, or when the deed was done. Like those Germans who, at the close of the Second World War, protested that they had not known about the extermination of the Jews and the Gypsies of Occupied Europe, we too are guilty of not having *looked*, not having *cared*, of having turned the other way *so we could not see*, thus hoping to be spared the shame of our collective deeds. But even if there were no shame, there would be a cost, a blood-payment, imposed for the extinction of large numbers of species with whom we share the planet, and ironically that cost may be imposed in plague and famine.

*　　*　　*

We are surrounded by life. It is under our feet, in the air we breathe, in the plants and animals we eat, in the water we drink and of which our bodies are, mostly, composed. If, through an electron microscope, we examine a gram of soil, a quantity so tiny as to be held in a crumb on our finger, we might find 10 billion microbes. In a teaspoonful of water — fresh or salt — we might count 1250 million viruses. We do not know what they are, or how many species, groups or families compose those millions. We know some of them: in the soil are the bacteria, for instance, that fix nitrogen from the air, and others that break down ordure, or other decaying matter such as the bodies from which the spark of life has departed, and separate from them the nutrients that can then be returned to life's great cycle. There will in the same gram be other microbes we know and fear: anthrax, listeria, salmonella and so on. These don't seem to exist in numbers which do us much harm, perhaps because the fierce competition for living space among 10 billion life forms within a gram of soil is itself a limit on the numbers of any single organism. Most of the microbes in water are still a mystery: some are agents of disease we have known for a century or more: cholera, for instance, and the polio virus; the others are a mystery simply because until 1989 we didn't know about them at all. Until the last few years, there has not been a tool that could magnify by the necessary 40,000 fold to see them, and until now we have used the tool to search for and study things specific: the known life forms that have worried or puzzled us; those to which we have needed an antidote or cure. Only in 1989 did anyone think of looking to see what, in total, was there in water taken from lakes and oceans, and only in that year was the astonishing prolificity of life in a teaspoonful — an unexplored universe in itself — revealed to us.

But then everywhere we look, we find life. We find it where our definitions of life say that it should not be. Our idea of life is based on the cycle of air, nutrients, water and sunlight, and our definitions of sterility, that is the absence of life, require toxic chemicals, extreme cold, darkness, cauterizing or scalding heat, or an excess of salt. And yet, a decade ago, at the bottom of the deep oceans, at tremendous pressures far removed from the light of the sun, in those fault lines where magma and boiling brines bubble up from the ocean crust, biologists discovered a different universe, a community of creatures who derived their energy from the heat of the water, and who consumed dissolved chemicals to build their cell structures, and on whom fed other creatures, and so on: a whole web of life removed from ours, and yet operating in parallel to it.

When the long dormant volcano of Mount St Helens erupted in Washington State, USA in 1980, it ejected a cubic mile of superheated debris into the air: a ton or so for every person on earth. Water temperatures in the rivers and lakes nearby soared towards boiling point. More than two million creatures — mammals, birds and fish — perished and trees and vegetation were vaporized, scorched or charred over an area 8 miles by 15 miles. The amount of energy released over a period of nine hours was roughly equivalent to 27,000 Hiroshimas. Yet scientists in flameproof suits were, within a year of the eruption, discovering life among the ashes: hundreds of millions of apparently flameproof bacteria rioting in the heat of every thimbleful of water, consuming ores and compounds of sulphur, iron, nickel, uranium and copper and using the energy of the reaction to build (against all our expectations of the word) flesh.

This obduracy, this drive of life to exist and even stubbornly flourish in any niche, hole or shelter, any environment available to it — at great temperatures and pressures deep in the earth's crust, at regions high and cold where the air is thin and the bombardment of ultraviolet is greater, in the most hostile deserts where it may not rain for years and the medium of shelter, the sand, constantly shifts and abrades in the rolling winds — creates problems for those who have tried to put numbers to creation's rollcall. In the first place, human vision is limited. It is not limited simply by our capacity to magnify or telescope, but also by our capacity to look. Until the last few decades, biologists thought that there might be as many as two million species of plant and animal on earth. Great museums such as the Natural History Museum in South Kensington and the Smithsonian in Washington hoped that at least between them they might eventually collect, identify, describe and preserve if not all of them, then nearly all. They didn't necessarily expect to complete the work. The fecundity of, for instance, the tropical rainforests had long since ceased to astonish them. The old joke about swinging a butterfly net and discovering at least two species hitherto unknown to science had long since ceased to be a joke. It kept happening.

Expeditions of entonomologists, botanists and zoologists would live in an area of the tropics, collect and preserve specimens and return to their laboratories to begin the long, sometimes tedious process of examining and describing them, and even before they had finished, two years later, they'd be off again, on a foray into another region, arranged long ago, and return with another series of undescribed, unidentified and unnamed creatures to be set in place and context, to be given an address and a rating valuation in Nature's electoral register. There would be other jokes, and apocryphal stories: of the crates that somebody found in a museum basement, which when opened turned out to be the collection of a long-dead Amazon explorer, deposited in the last century, and never actually examined, and which *still*, all that time later, turned out to contain 120 species hitherto unknown to science. Some museum discoveries were not jokes. One beetle was actually discovered, examined, and given a name, and declared a serious pest to be exterminated ruthlessly wherever found, all inside a museum. It is now called the museum beetle. It lives on the bodies of long dead insects and can obliterate whole museum collections of insect specimens. So far, about one million species have been named and catalogued. About one new mammal and about three species of birds have been named and catalogued each year. But mammals and birds, even the smallest, are quite big. Elephants, whales, mice, men and women, storks, humming-birds, and even bees are all easily visible. But about 99 per cent of creation is less than 3mm in length. Furthermore, we have not been looking everywhere. The temperate zones — which include the nations with scientific support for agri-culture, huge chemical industries, hygienic cities, universities and a working population with an interest in and enthusiasm for fishing, bird-watching, butterfly collecting and wildflower pressing, rambling, picnicking, and the preservation of national parks, are and have been for more than 100 years under continuous study from a huge army of professional and talented amateur naturalists. A tropical nation like Panama, or Colombia, or Indonesia has no more than a handful of trained botanists, foresters or zoologists. And yet (to choose an example at random) there are more species of wildflower in the tiny state of Panama alone than there are in the whole of Europe.

When biologists tried to guess the number of species in the world they took the species counts of the temperate zones and multiplied by three, guessing the tropics to be at least three times richer in all types of plant, insect, mammal and bird. A very crude calculation like this, and a growing awareness that most of the species of the world had yet to be described and named, allowed biologists to think that there might be more than two million species: perhaps three million, perhaps five.

But even tropical biologists have limited vision: limited, that is, by how far they can see. The richest habitats in the world are the tropical rainforests. These are dark, enclosed, sheathed by the foliage of trees that grow to 35 or 40 metres. The men with the bush knives, butterfly nets and collecting boxes can only observe the understorey, the lower branches and the tangles of life in the trunks and roots, and on the river banks. Once they found techniques for studying the forest canopy — where the sunshine is, and the flowers and fruits and green leaves wet with the first drops of rain — they were startled to find yet another unexplored world.

The most famous study was of one particular evergreen species in Panama. A decade ago, an entomologist painstakingly collected all the beetles he could from the upper branches, and began sorting them. He found an astonishing 1100 species of canopy dwelling beetles — carnivores, herbivores, fungus-eaters, scavengers — on that one species. He calculated that a proportion of each class were host-specific: that is, roughly 160 of those canopy-dwelling beetles could only live in that one species of tree. Beetles are a class of creature known as arthrophods: as far as we know, they make up 40 per cent of all known arthropods. Therefore, he argued, there would be 400 arthropods that could only live in that species of tree. Then he suggested that canopy arthropods would be twice as rich as forest-floor arthropods: that made a total of 600 arthropods per tropical tree species. But there are 50,000 known species of tropical tree. Therefore, there might be a total of 30 million tropical arthropods.

The sums are crude. There is no particular evidence that the bases of his calculations are right or wrong. The answer could as easily be 15 million or 65 million. The point is that his study showed, quite clearly, that the scientific laboratories and museums of the world simply do not have the resources, the time, or the trained personnel to identify, classify, or study them all. And, of course, we have been dealing only with arthropods. Each arthropod must have its own suite of parasites, bacteria and viruses dependent upon it, in the way that the arthropod depends upon the tree.

Once again, the numbers threaten to become dizzying: wherever we look, there is life, and further life depending upon that life, and yet other creatures who live in symbiosis with that life. But it is worth remembering that this astonishing figure of 30 million species of arthropod is based on one study of one particular tree in one country in one climate. It is also worth remembering that that tree, and all the others around it are being felled, cleared and burned, the land sown with grass and then sprayed with pesticides, on an unbelievable scale: an area the size of Belgium, every year. And most zoologists agree: if the rate of felling of the tropical forests continues, or continues to increase, then in the next 50 to 100 years more than half of all the species with which we share this planet will become extinct: they will disappear forever, and we will not even have known that they were there.

We are living in an era of mass extinction. It is not a new process, but a more violent one. Humans have been making other creatures extinct from the earliest days. We eliminated the wolf and the bear from Britain; but the wolf and bear continued to exist, by retreating deeper into the remaining wildernesses of Europe and Asia. Creatures that exist only on small islands, however, cannot retreat: thus the dodo disappeared from its Indian Island home quite quickly after its discovery; the first Maori settlers in New Zealand swiftly eliminated the giant moa. Humans are predators: like all predators we drive away those other predators that compete with us for food, and like all predators we prey on slow moving, edible herbivores, or creatures which have some economic value to, and which are at the same time easy targets. But precisely because of this, precisely because they are big, unmissable, you can also see them going.

That is why the tiger and the elephant and the whale and the mountain gorilla are all at risk of extinction, and why there are international programmes to try to save these species; that, too, is why the two Superpowers, during the height of the Cold War, managed to push through an agreement to protect the polar bear; why both American and European bison have been rescued from the brink of extinction; why there are multi-nation projects to save the desert oryx, Pere David's deer, the wild horse from the Asian steppes; and why the giant panda from the bamboo forests of China has become the symbol everywhere of international concern for the environment. Smaller, more specialized creatures aren't so easy to see. The notornis, a largish, shy flightless bird of the New Zealand marshlands was believed to be extinct. Magically a pair was rediscovered in the 1950s: the species is now under expensive, permanent surveillance. The notornis survives, but its survival will always be precarious.

Until the arrival of humans, there were almost no mammals on New Zealand. There, flightless birds appeared to take on the roles of mammals in other environments: they lived in holes in the ground, grazed in flocks, sheltered under bushes from predatory hawks. Even the vegetation took a different aspect. With the arrival of settlers — and deer, goats, pigs, opossums, starlings, dogs and cats — the native plants and animals were at immediate risk. Herbivores that grubbed among the roots, that ate the tips of new shoots, that stripped bark and chewed branches, began to change the nature of the vegetation, and even began the process of stripping whole hillsides. Feral cats and dogs took birds in the way they might, in Europe, have taken mice. The process of elimination was unconscious, unintended, unobserved and inexorable. And it continues, despite the best efforts of a nation which takes its own flora and fauna seriously: in one horrendous episode, one feral Alsatian dog killed 500 of the 900 brown kiwis in the Waitangi state forest of the New Zealand northland. The kiwi is New Zealand's national bird. The lesson of the notornis and the kiwi is clear: if they are to survive they can only survive under guard. They will no longer be wild things, but dependents of the human race.

Some, of course, don't survive. When the US space agency Nasa drained the swamps of Cape Canaveral to build a launch pad for the moon, they altered the habitat of the Florida dusky seaside sparrow, and a perky, rather drab little species with a restricted habitat began its swift dive into oblivion. There was, however, immediate consternation, and a massive million dollar effort to save the little creature. Six sparrows were, with great difficulty, caught and established in a breeding colony. There was only one problem. They were all males. They waited, forlornly for even one female to be caught and put with them, and they waited in vain.

These are just a few individual stories in a swelling chorus of mass tragedy. There are about 1000 mammal species heading towards extinction. Practically all the big cats — leopards, jaguars, tigers, ocelots, cervals, cougars, pumas and so on — are about to steal away into the endless night. Ten species of bear are about to go; 23 species of whale; 15 species of seal. The chimpanzee is at risk: it is used for research into the disease of Aids, and for every live chimpanzee brought to the laboratories perhaps 10 have died in capture or transport. The African elephant is being destroyed by poachers, to the execration of the world, which sees the destruction bloodily on television, but cannot stop the slaughter. Some smaller creatures had all but disappeared before we really knew they were there: the Darling Downs hopping mouse and the lesser rabbit-eared bandicoot of Australia; and the groove-toothed mouse of Togo, observed once in 1890 and then never again afterwards. There are only 9000 species of birds in the world: more than 1000 of them may be about to fly away forever: farewell the Chinese crested tern, the Algerian nuthatch, Pallas's fish eagle, the seven-coloured tanager, the West Indian whistling duck, the Madagascar pochard, the Mauritius cuckoo-shrike, the hyacinth macaw and the Grenada dove, the resplendent quetzal (Guatemala's national bird), the imperial eagle which ranges from Europe to Pakistan, and is ranging more thinly everywhere.

The only surviving Californian condors now live in cages. This may be the fate of many more species on their last legs; to see out their days on perches, in a series of sometimes successful, sometimes futile bids to prevent the last of their kinds from disappearing for ever. These attempts may indeed be futile: those agencies which have answered the challenges inherent in species rescues have found the going tough. In some cases the animals have responded to captive breeding and have been released again into dwindling habitats on the understanding that they and their habitat will thereafter be protected. Thus the golden lion tamarin is back again in the forests of the Atlantic coast of Brazil, although more than 98 per cent of these beautiful forests have been destroyed, and the scimitar-horned oryx is back in the deserts of the Middle East.

But what price the salvation of the Siberian tiger? There are about 200 left in the wild, perhaps fewer. Since even those hostile regions are steadily being cleared and settled, and since the available prey is itself dwindling, each tiger needs a larger and larger range to feed itself, and the chances of pairing and survival shrink accordingly. The Siberian tiger probably doesn't stand a chance in the wild. On the other hand, they can now be bred quite successfully in zoos. But the serious zoos, the ones devoted to zoological and conservation research, can probably house no more than 500 altogether. Siberian tigers can have two or three litters a year. If each pair has only one litter every two years then the zoo population of Siberian tigers — each large, expensive to keep, needing an area to prowl, and a diet of high quality meat — could soar in two decades to 6500. This means that zoos will have to shoot 300 tigers a year just to keep a core population alive, and may go on having to do this for 200 years, until the global population falls, and some form of international instrument can be devised to recreate large areas of genuine wilderness, that is territory in which man need not interfere and in which prey and predator can keep themselves in balance, and into which the tigers can be released. But at the end of this cruel paradox is another. The tigers that have flourished in the zoos will not be like those that might have flourished in the wild: natural selection is just that, a selection for survival in the habitat to hand. The animals that emerge at the end of 200 years will have been selected for their capacity to gain weight in Berlin or San Diego or Regent's Park, and keep their tempers in cages and moated territories, and to appreciate the delivery of haunches of beef every second day. How much at home will they be in the taiga?

There are even crueller paradoxes: the Siberian tiger is only one of 1000 threatened mammals and 1000 threatened birds, and the processes of mass extinction will not stop when those thousands are gone. To save any of them means saving a population of about 500 — think about it: could one be said to 'save' the colossal variety of the human race by selecting 10, or 20, or 50 people from say, Winnetka, Illinois or Chelten-ham Spa in England? Especially if some of them carried genes for inheritable illness such as sickle cell anaemia or Huntingdon's chorea, or carried the Aids virus. That's why a large number like 500 is necessary. But to save 500 specimens each of 2000 mammals for even 20 years would cost the world's great zoos about as much as it cost the Americans to put a man on the moon. And the need for salvation will go on for longer than 20 years: all the indications are that the rates of extinction are going to get worse.

Between the years 1600 and 1900, humans hunted to extinction, or crowded out, roughly one species every four years. During this century, known extinctions have been occurring at the rate of about one a year. It's a matter of roads, and car parks, and quarries, and drained swamps, cleared scrubland, ploughed prairie, overgrazed savannah, felled forest, diverted rivers, industrial outfall into estuaries, new town development, sprayed crops, and factory pollution: very little of it now is simple overhunting or trade in skins and household pets. The rate is increasing. Right now, at the time of writing, it is put at about one extinction per hour. (We owe these calculations to Professor Norman Myers, the tropical forests expert, and his book *The Sinking Ark*: they are now widely accepted simply because, even if they are wrong in detail, they are not wrong in the final sum, unless they are too low.)

By the end of the century, the toll could be horrendous, with each tick of the clock marking another creature's passage into oblivion: possibly a million species eliminated in the last 25 years of the millennium, at the rate of 40,000 a year. For decades now, the world's conservationists have been recording threatened species in a series of Red Data Books: these have grown thicker and greater in number, but those tellers of animal mortality cannot keep up. There is already a proposal that, instead of Red Data Books, there should simply be Green Books which list those species which are known not to be at risk: and everything else should be regarded as endangered. This is because some creatures are almost certainly extinct even before they can be classed by the conservationist bureaucracy as vulnerable. And behind the creatures being entered, one by one, in the ledgers of threat, the other candidates for oblivion pile up. The USA announced in 1989 that 680 US plant species will be extinct in the wild by the year 2000: that is, 680 species they *know* about. That does not, of course, include those creatures whose life cycles depend on those plants, nor the sometimes fragile networks of other animal and vegetable life in which each of those plants provides a link.

Nor are those 680 plants the only plants known to be at risk. There are between 250,000 and 300,000 flowering plants in the world. Altogether about 25,000 of them are threatened, endangered or perhaps already extinct. Plants, like animals, often have restricted habitats. In one case a botanist catalogued the flora of a mountain ridge in Ecuador: on it, he found 100 species of plant endemic to that ridge and occurring nowhere else. Within four years the same ridge had been completely cleared for agriculture. That is, in four years, the same biologist who mapped 100 species watched them disappear forever. At least someone saw them go. Almost certainly, the same rate of destruction is occurring in Amazonia, in the other Andean States, in Central America, in forested Africa, in the Indonesian and Malay archipelagos, and the forests of India and China and there is no recording eye to see their passing. And as the plants go, so do the animals that rely upon them for life, the insects and the birds that prey on the insects or the nectar or the seeds.

The gloomiest view — which has an ugly ring of possibility to it — is that in the end humans will share the world with those creatures of known and tested economic value to them — but we have yet to come to the question of economic value — sheep, goats, cattle, pigs and so on, all of them far removed by centuries of crossbreeding from the wild stock from which they came, and the only creatures of the wild that survive will be those of no economic value and no great nuisance, but which will be most like humans. They will be fecund, opportunistic, resourceful, scavenging, not frightened of humans, omnivorous, prepared to survive in cities, and ready to make their homes anywhere. It will be odd sharing the globe only with species like the cockroach and the Norway rat, the starling and the sparrow, the thistle and the Oxford ragwort.

CHAPTER 16

Metropolis of root and branch

In a city, things exist which could not exist in a small rural village or hamlet. A village dweller cut off from the towns must be a bit of a gardener, farmer, forester, shepherd, hunter, haymaker, thatcher or well-digger, and have the skills to repair a home and preserve food. The collective surplus of the village may also support a teacher, a clergyman, a shopkeeper, a mechanic or a blacksmith: not much more.

Multiply the population of a village a millionfold or so and something quite different exists. There appear specialized creatures who provide not food so much as mortadella, mozzarella and olive oil; or knackwurst, bratwurst and pumpernickel; there are others who handle only money — that is, the tokens of primary production, the symbols of wealth; and yet others who, at one remove, handle only the numbering of money, and people beyond them who parasitize off the traffic of currency (itself a traffic merely of electronic numerals on a screen) from one nation to another; there are some who sell cars and others who survive by directing the traffic of cars and a third group who exact a toll where the car comes to rest and a fourth who scavenge off the rusting bodies of exhausted automobiles. There are mechanics who deal only in the repairs of old Citroen models; some who sell only Volkswagens and some who trade only in copies of VW insignia as badges for the young; there are businesses devoted to the cutting of gemstones and yet more businesses which provide the silver fixings and yet more businesses to sell them and a fourth tier of enterprises which buy them or hold them in pawn. There are people who make false stone facings for buildings already clad in brick; there are people who bring rubber plants into offices, call to water and feed them and then take them away again and replace them with weeping figs; and there are others who sell not intoxicating liquor but wine to be drunk slowly and with respect; or who dance, and then teach dancing; or who travel from apartment to apartment with the equipment for setting hair, clipping fingernails or providing colonnic irrigation; or there are specialists who trade on the streets: they sell heroin, tins of London fog, cocaine, bagels, amphetamines, ice cream, hot dogs, tee shirts that say 'I love New York', or their bodies, in ways not to be imagined in a small community; there are others who arrange house sales, or theatre tickets, or sex-change operations, or who devote enormous energy and imagination to devising posters to sell cigarettes without either showing or even mentioning cigarettes, or who form small but flourishing consortia for the manufacture of uncomfortable latex dresses for men who want to look like women; there are small and even large concerns devoted entirely to the care of the detritus of the past, and others dedicated to speculating upon the imponderables of the future. And all of these specializations have evolved in symbiosis with each other, all, without exception, are making a living off the needs of each other: preying, devouring, parasitizing, scavenging; filling niches in the economic structure erected within a city; and the bigger the city, the more varied and more precisely tailored the number of niches available to be filled. Because each specialization branches out, and each creates more opportunities for more hangers on, more specialized parasites, more selective scavengers, more finicking predators.

But then, quite suddenly, and almost unconsciously, we have stopped talking about a city at all: we are talking about something else, something like a tropical mangrove swamp, a coral reef, or a tropical moist forest. Before we leave the parallel of the city entirely, there are a few more points to the analogy. One is that a city is an accident of geography lost in history. It is, unless the favouring circumstances change dramatically — unless it is buried in desert sands or swept away by the sea or entombed in volcanic lava or sacked and its population butchered or sold into slavery and its very fabric levelled and its soil sown with salt — a self-sustaining entity: its own mass attracts and fosters new specializations as old ones are lost. So, too, is a rainforest self-sustaining: creating and maintaining to a certain extent its own climate;

providing for the creatures within it the raw materials and currency for life from its own detritus and decay. Its economy is fed by the sun, but it guarantees much of its own rainfall in a way that the city guarantees much of its own wealth, and its denizens live, as do those of a city, by taking their cut as the currency circulates, and passing it on again. There is another parallel. Both entities are examples of huge, interlocking webs of interdependencies. There is a third: in a forest, as in a city, there is a vast, ordered but not cross-indexed, pool of knowledge and resource.

The fourth is a more sombre parallel. In a city, though the number of specializations is enormous, the number practising within each specialization is sometimes quite small, and the circumstances that permit their survival are often precarious, subject to seasonal change and occasional collapse. Even the greatest city can support only so many cat-burglars, so many petrodollar lawyers, so many bootleggers, heroin assayers, hairdressers, arthropod taxonomists or contract assassins. This is true, too, of the even greater, richer and more precise specializations within the tropical forests. But there the analogy collapses. When circumstances change, mineral water salesmen and cat-burglars can retrain as bakers or brain surgeons. A fig-wasp that lives in symbiosis with one and only one species of fig cannot survive when the fig has gone. Nor, if the wasp should go first, can the fig.

The rainforests cover no more than 7 per cent of the Earth's surface. But they are home to more than 50 per cent of the Earth's species. There are about 9 million square kilometres of rainforests left. In total, this is an area equivalent to that of the USA. But these forests are being cleared at a rate of 200,000 square kilometres — an area bigger than the UK — each year. Between 70 and 90 per cent of all the extinctions taking place today are taking place inside the rainforests. Right now, at the time of writing, the obliteration of three isolated rainforests is almost complete. These are the rainforests of Madagascar, of the Pacific coast of Ecuador, and of the Atlantic coast of Brazil. Within these clearances, around 100,000 species will have been destroyed.

By the end of the century, less than a decade from now, the rainforests of Central America, Southeast Asia, West Africa, the foothills of the Himalayas and the Pacific Islands will have all but gone, and with them a million square kilometres of trees and half a million species. In 50 or 60 years — within the lifetime of a child now at school — the only large rainforests left in the world will be in the Congo basin, and in Western Amazonia. Once again, we are confronted with the peculiar blankness, the numbness of sensibility, that accompanies figures such as 'half a million species' and 'a million square kilometres'. A million square kilometres is a felling and burning of an area as big as Burma and Burkina Faso combined; an area once thickly and richly layered with natural life, the size of Egypt, rendered as sterile as, well Egypt. Half a million species, too, is an unimaginable figure. For comparison, and if it helps, biologists have, in the 250 years since the work of Linnaeus who gave his name to the method of systematic classification and naming of creatures, managed to name and classify only about 1.4 million species of flowering plant, fungus, mammal, bird and insect.

Nor is there any serious hope that, on any scale, species can be saved by the creation of reserves: isolated stands of forest, protected, as are national parks and sites of special scientific interest, in Europe and America. (Imagine trying to 'preserve' Paris, London and New York by preserving only St Cloud, Earls Court and Yonkers and bulldozing and then ploughing and sowing ryegrass on the rest of those cities.) There could be several separate, but interlocking reasons for this. The large predators, for instance, require big territories: if the preserved area is not big enough to hold a viable breeding population, they will die out. If the predators die out, the prey multiplies, and exhausts its own food supply: extinctions will cascade anyway as whole chains of predator, prey, food sources and the parasites dependent on all three perish.

Then there is the problem of variety: a vast forest holds a number of separate habitats; a small forest may hold only one. But many plant and animal species depend on two or more habitats — the pollinators and seed spreaders, for instance, migrate between ecological zones according to the season. A quarter of all the tree species in the Barro Colorado reserve of Panama are believed to be 'subsidized' in this way by the forests around the reserve. When the surrounding forests have gone, the reserve will swiftly become the poorer. A third reason is that, in forests as in cities and nations, sheer size is strength: the predators and parasites from 'outside' can penetrate deeper into a small forest than into a vast one. A jungle the size of the Amazon basin can survive being, as it were, nibbled at the edges because the variety of its inner life remains undisturbed, and flourishes to renew its perimeters, simply because the ratio of boundary to area is smaller. In a small reserve, with a larger ratio under menace, the feral dog and cat, butterfly collector and parrot trapper, orchid hunter and the destroying insects from the savannahs can do proportionately more harm. And each wound in the fabric of the forest's life multiplies: each extinction is followed by a chain of other extinctions.

Nor is there any serious hope that the forest can re-establish itself, to its former richness, by 'growing outwards' over derelict farmland or rubber plantation from a preserved area of wilderness. It might do so — although probably not in the peculiar case of the Amazon — but it would remain impoverished. Once again, the analogy of the city is helpful. If you reduced London to Earl's Court for a few hundred years and then began to allow it to grow, there would in the end, once again be banks, estate agents, police stations, insurance offices, and bookmaker's offices, supermarkets and video rental stores from Ealing to Hackney. But would anybody bother to re-establish the College of Heralds? The Chelsea Pensioners? The Pearly King? The Beefeaters and the Yeomen of the Guard? The Karl Marx Memorial Library? The Chelsea Physic Garden? Or for that matter the Victoria and Albert Museum?

There is another complication. It is to do with the food supplies in a forest: particularly the fruits on which primates and birds rely. In the Manu National Park in the Peruvian rainforest there are nine months in which the trees provide far more fruit than the fruit eaters can consume. But for three months, in the dry season, 80 per cent of the 'biomass', that is, 80 per cent by weight of the animals of the forest, must survive on the fruits — the palm nuts, figs and nectar — of just 10 species of tree. Those trees are almost certainly thinly spread.

The thing no reader who has not seen a tropical forest can ever quite imagine is the sheer variety of the canopy. In a single, one-hectare plot in Peru, one botanist counted 600 trees, divided into an astonishing 300 species. The chance of any particular plot having more than one example of any of those 10 keystone species is quite small. It may not even have

The water cycle

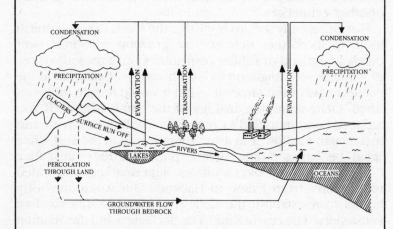

Most of the world's water is in the oceans, too saline for use. Most of the globe's fresh water is in the ice-caps. At any time only a small percentage is in the atmosphere (eventually to fall as snow, rain, hail or dew) or in the lakes and rivers. The estimate is that there are only about 9000 cubic kilometres of freshwater available at any one time in the world, and humans are using a third or more of that already for industry and agriculture. About one sixth of this fresh water is in the Amazon, which is one reason why there is worldwide concern about the destruction of the rainforests there.

one of them. Those 10 species are less than 1 per cent of the plant species of the whole forest. Thus, although the forest is teeming with fruitbearing trees, four-fifths of the vast mass of life, along with those predators and parasites who rely on the fruit eaters, is utterly dependent on less than 1 per cent of the forest species. Any reserve which contained too few of these 'keystone' species would itself be useless: without the keystones, the edifice of life erected upon them would crumble. You cannot hope to preserve the Empire State Building when its ground floor has been demolished.

There is another reason: it overrides all others, and it applies more to the Amazon than perhaps any other forest. The rainforest and the rain go together. If one goes, so might the other. A large proportion of the torrential downpours that water the Amazon basin — perhaps 50 per cent or more — is actually manipulated by the forest itself. About half of the annual rainfall blows in from the Atlantic, and much of this falls on the slopes of the Andes. The rest is provided by the forest itself.

The Amazon canopy is its own climate machine: being deep green it has a low albedo, absorbing radiation from the sun in great quantities, slowing the rate of evaporation of water from the soil although the trees in fact have to use enormous quantities of water, just to stop themselves overheating. But even this works to the forest's advantage: the water transpired from the trees builds up over the forest; air currents collect it into clouds; the clouds condense into rain again, and release once again the water that may have evaporated from the cooling trees only the day before. Thus the forest, if big enough, is like a huge pump, continuously showering and refreshing itself with the same water. Even so, and despite the enormous quantities needed to cool and nourish itself, the foliage, roots, underbrush and humus of the forest floor retain vast lakes of yet more water, releasing it into the Amazon river system throughout the year, including the dry months. Since the catchment area of the Amazon is about 6 million square kilometres, and since about one sixth of all the river water in the world passes through the Amazon itself, the forest is quite clearly not just important to its own management, but important to the climate machine of the whole world.

But outside the forest, things are different. The cleared land within and around it holds almost no water: it runs off, leaching the soil swiftly of its nutrients: or the water evaporates more swiftly and the barer soils reflect the sun's radiation. There is therefore an air temperature difference between the two. This difference can be considerable. In 1989 satellite scientists began examining the cost of overgrazing. They took measurements of the temperatures across the border between Texas (where there is legislation to limit the number of grazing cattle per area) and Mexico, where there isn't. They found a distinct temperature boundary line. The temperature on the Mexican side of the fence was 4°C higher than on the Texas side. And that — the temperature difference between a large city and the countryside around it — is the difference between dry plains and semi-arid plains. The difference between a dense moist forest and the bare rangeland around it would be even greater. If the rangeland was large and the forest was small the difference would be overwhelming. The forest would parch and fade anyway and with it those species that remained. That is why the remaining Amazon forest and the creatures within it cannot be 'protected' by preserving selected blocks of it. In the case of a rainforest, survival and scale are not separable.

So far, we have been contemplating the survival of the rainforest, and particularly the Amazon rainforest, for its own sake, or for the sake of the species to which it plays host, or for the sake of the tribes of forest dwellers within it. There are economic reasons as well. They have been well-rehearsed, but they are worth repeating, in what is, in effect, a geographical handbook of calamity. The Amazon forest may be richer in life than any other habitat on the globe, but the richness is precarious. The soils in which the forest is rooted are surprisingly poor. They have been leached and weathered and pounded by heat over millions of years and most of the nutrient minerals in the soil have been washed away. It happens in other parts of the world, too: nutrients are washed away or blown away in the eroding winds, but they are replaced by the weathering of the subsoil beneath or by fresh river silt at flood time.

The Amazon basin is, largely, different. The subsoil is poor, too. The nutrients have been all but exhausted. The apparent inexhaustibility, the fecundity and vitality of the forest is illusory. The trees themselves, some of them vast monuments to Nature's injunction to reach for the sky, to find a place in the sun, to stand head and shoulders above the rest, are only shallow rooted. Their roots do not explore the deep soil for nourishment and water: there is none, so they take it from the surface. They survive on the excrement of the animals that eat their fruit, and the compost of rotting foliage on the surface below them, and on the dissolved nutrients in the raindrops that drip from their own leaves to the surface below them. The roots are near the surface precisely so that the trees can once again tap the nutrient minerals and nitrogen from their own debris, get the best of it before the stuff vital to their own survival seeps below ground into the water table, and soaks away into the rivers and is carried out, ultimately to sea.

The forest Indians may clear part of the forest for their own crops, but within a year or three the soil is exhausted, and they move on. The damage to the forest is tiny: it may even help the forest, by providing occasional clearings that can be colonized by those smaller plants that need light, and are denied it in the older stretches of woodland. There is a net loss of nutrient to the economy of the forest, but it is small, and the forest is large, and can shrug off the loss, just as any successful restaurant can survive the occasional diner who departs without paying his bill, or any city can survive the collapse of one industry.

However, the wholesale clearing of a forest, on a scale involving hundreds of square miles, is a different matter. The bedrock of the Amazon region, more particularly of the Guyanan and Brazilian shields to the north and south, is among the oldest known to us. It is called Precambrian: our name for the oldest rock of all, rock so old it carries almost no fossil evidence of life at all. This rock has weathered and eroded into a kaolinitic clay which has collected in the Amazon basin, and then been weathered and eroded again. The magic of time, water and sunlight and the evolutionary goad has somehow erected upon the clay the vast cathedral of the Amazon canopy, but it is a cathedral which, if it toppled, might never be erected again.

The felling of a 100,000 acres or so yields an immediate income in tropical timber, and the burning of the underbrush and the useless trees and foliage covers the ground with a bonus of immediate nutrient and for a few years the cleared land shows promise as a cattle ranch, or cash-crop plantation.

But a clearing of starved clay that receives three or even six metres of rain per year cannot stay productive for long. The first heavy rains begin the process of dissolving and carrying away the nitrogen, phosphorus and potash but while these minerals last, weeds will spring up. They may be such as to need burning and ploughing again, but by this time the dry weather will have set in and the soil will be after many thousands of years, exposed to the fierce glare of the sun. Disaster doesn't become apparent in the first year. But — except in those occasional patches of Amazon basin soils where the nutrients have, by some accident of geology, accreted a little rather than been lost — it arrives sooner or later. The technical term is laterization. Take some ground made of kaolinitic clay: wet it, tread it down, dry it, wet it again, pound it, dry it again and it turns into laterite. It sets hard in the end and stays hard. Trees can't colonize it: the jungle has been neutralized. The laterite sets so hard that the rain won't penetrate it at all, but runs off instead in great sheets, leaving a baked dry soil underneath.

In the Amazon basin, cement and asphalt are difficult to come by: they have to be hauled long distances and they deteriorate in the humidity. So the developers surface roads and airstrips with laterite. They get the laterite from abandoned homesteads and spent plantations and deserted ranches: ranches that once meant hope to families and which now spell despair and economic ruin. And the laterite from those cleared lands is then invested in making roads deeper into the forest, to repeat the process. And for that, for nothing, the greatest forest in the globe is being destroyed and millions of years of evolutionary adaptation being obliterated every hour: just so that we can actually make a bit of money for two or three years and then leave behind a desert: a most unusual desert, one with a surface like a tennis court, one which gets two or three metres of rain a year; a desert that, over an area big enough for a British or North American house and garden, might have maintained, as seedling, sapling or giant, not 265 trees but 265 *species* of tree; a desert that might, over a few square miles, have maintained more species of flowering plant than there are in Europe, and with them such a multiplicity of parasites, creepers, epiphytes, rodents, birds, beetles, ants, fungi, butterflies, moths, and reptiles that all the biologists in the world could not find the time to catalogue them.

Not, of course, that they will get the time to catalogue them. Not at the rate that the forests are disappearing. In the time it took you to read this chapter, at least 11 square kilometres of tropical rainforest was destroyed forever, and another vast gulp of carbon dioxide delivered into the atmosphere to fuel the greenhouse effect. And the only economic benefit from it will be a few extra hamburgers, a few crates of trademarked cola, a few jars of instant coffee, some hardwood furniture you don't really need, and a very large tennis court. On which, of course, nobody can play. And after a while, as the jungle recedes and the tennis court gets bigger, the rains, too, will stop falling. Because, of course, the rain and the rainforest go together, and when one goes, so, after a while, does the other. It doesn't seem much of a bargain. And anyway, leaving aside the greenhouse effect, our own potential economic losses go much deeper.

CHAPTER 17

The fruits of the forest

In 1989, three US botanists took a close look at the economics of human action in the Amazon rainforest. They chose a small riverside village on the Rio Nanay, southwest of Iquitos in Peru where the local residents made a living from shifting cultivation, fishing and collecting forest products for sale in Iquitos. They marked out a one hectare area of the forest and then took a botanical inventory. They found 824 trees of 10 centimetres or larger in diameter. Those trees were of 275 species. Of these, 72 species yielded products that could be sold in a market, and that one hectare contained 350 of these economically exploitable trees. That is, in one hectare, more than 41 per cent of all the trees were actually used for monetary gain. Eleven species bore edible fruits, 60 yielded commercial timber and one of them could be tapped for rubber. The same region also contained medicinal plants, lianas and palms that could be used commercially, or were important directly to the collectors, but the researchers didn't include them in the tally. They checked prices in the markets and the local sawmills, they counted and weighed the fruits, they calculated the time spent on harvesting in the forests and then deducted costs at the minimum Peruvian wage rate, together with the expense of selective logging. They made allowances for the fruits that wouldn't be taken, to allow plants to regenerate. Then they took a look at costs, inputs and yields from other parts of the rainforest, parts where the forest was being clear-felled and planted, and parts where cleared forest was already being farmed. Then they used a standard economic technique to match the yield from forest harvesting against the yield from plantation logging and from ranching over similar one hectare plots. They included in their calculations not just present but future yields from uses of a hectare of Amazon land, forested and cleared. They then arrived at a measure of economic worth which is called the net present value.

The net present value of the hectare of forest, just in terms of the potential in fruit and latex, was $6,330. The value of the timber on the hectare was about $1,000, delivered to the sawmill, but one clear-felling would have destroyed many of the fruit trees as well. On the other hand, if valuable timber trees were taken selectively on a sustainable basis, the net present value of the forest rose to $6,820.

A hectare of Brazilian forest, converted to a plantation of one species for timber and pulpwood, and subjected to the same techniques of study, on the other hand, had a net present value of less than half that figure: $3,184. The net present value from fully-stocked cattle pastures in Brazil had a value of only $2,960: less if you deduct the costs of weeding, fencing and animal care. And in making these calculations the three scientists made the assumption that both plantation forestry and cattle grazing were sustainable forms of land use in the tropics, even though they thought this was an unduly optimistic assumption.

The message to the governments of Peru and Brazil ought to have been quite clear. The economic value of the forest to the people who live in it, and thus to the nation as a whole, is far greater if the forest is left more or less as it is. The difference is, of course, that tropical timber and pastured beef earn foreign exchange, and the value of them is entered into national accounts by people in business suits. The resources of the forest, taken to the markets in basketfuls by men and women in home-made sandals, might as well be in the black economy: nevertheless, they provide a securer livelihood and a greater spread of incomes for a large number of people, and the forests will go on providing that livelihood for as long as they are permitted to stand.

What is more, there are some forest products that, until recently at least, only the forest could provide. One of them is that European favourite, the Brazil nut, whose life cycle sounds like an obstacle race, and looks like a perfect example of the unexpected webs of interdependency on which the natural economy of the forest rests. Its flower has a hood to protect both nectar and pollen and its seeds are protected by a shell which can, in the wild, only be opened by an agouti, a forest rodent with powerful jaws. It has been harvested entirely from the wild, by forest residents who have left the forests standing. This is because it couldn't be planted and farmed: until recently no one knew how it was pollinated.

We now know that there are only a few species of bee strong enough to get into the flower. The most important of these bees has a mating cycle which depends on an orchid: the males gather fragrances from it and use it in a group mating dance to which females are attracted. Thus the male bee is important to the orchid — because he carries the pollen from one flower to another — and the orchid to the bee. The female bee, however, is important to the Brazil nut: only she can prise open the flower to take the nectar, cross-pollinate and thus set the seed. When the nuts ripen and fall, the agouti chews through the hard woody fruit, as big as a baseball, removes the nuts, eats some and buries the rest in caches. Sometimes he remembers where the caches are, sometimes he doesn't. It's Nature's way of making sure the Brazil nut pops up everywhere in the forest. So we have a wild tree which depends directly on a bee, an orchid and a rodent, and a worldwide commerce that has, up till now depended on the skill and knowledge of the peoples prepared to live with the forest, rather than destroy it.

The message couldn't be clearer. The forest knows what it is doing. The forest tribes and the small settlements on the riverbanks can benefit indefinitely from the forests: they could provide greater riches without serious loss to the creatures of the Amazon if the trade in exotic fruits and nuts and latex were expanded. On the other hand nobody, in the long run, benefits from forest destruction. But the economic reasons for the maintenance of the rainforests go deeper, and involve investments of millions of pounds or dollars, and potential or existing commercial markets of many billions. They are these. Within the forests — and indeed within plants everywhere — lie the future of our pharmaceutical industries, the future of some of our industrial and energy supplies, and almost certainly the future of our food supplies. Quite literally, there is within the rainforests the wherewithal to protect many of us against plague and famine, and many of our crops against blight and failure.

First, plague, illness, sickness, suffering and death. We are all creatures of herbal medicine. We may imagine that modern medicine is the gift of modern science; that drugs are designed and prepared in laboratories and produced in large factories. We are right in a sense. Most drugs are synthesized in laboratories called plants and produced in large factories called forests. The only part of the gift from modern science and industry is the recognition of these drugs, their extraction and preparation on a huge scale, and their dispensation in small, labelled brown bottles. Most of us know about aspirin, which was first prepared from willow bark, and digitalis, from the foxglove, and the antimalarial drug quinine from the South American cinchona tree; most of us by now have heard of the Madagascar rosy periwinkle which in the last decade or so was discovered to contain not one but two powerful alkaloids: the one effective against Hodgkin's disease, the other against acute lymphocytic leukaemia. The commercial income from these two substances alone is worth $100 million a year, and who would put a value on the relief of suffering and misery that grew in one small, attractive, but not otherwise very distinguished plant?

The rosy periwinkle is, by the way, probably heading towards extinction in the wild, its original home. There are five other species of the same family in the same territory. None of them have yet been studied. If they disappear in the wild — and this may happen, because the forests of Madagascar are disappearing — we will have lost not just a potential quiver of medicament treatments for disease. We will have lost the habitat, that forcing house of evolutionary pressures that triggered the rosy periwinkle and its cousins into producing these alkaloids in the first place. And this would be a pity, because the better we understand why plants produce the things they do, the more likely we are to know what to look for when we go searching for a better pharmaceutical treatment for Aids, or some of the cancers, or influenza, or cerebral malaria.

But aspirin, periwinkle and quinine are not isolated examples. Substances from plants provide us with painkillers, antibiotics, heart drugs and hormones and oral contraceptives, laxatives and cough medicines and pesticides and things that stop blood clotting. Most of these are not synthesized in pharmaceutical factories from the mineral template or by genetically engineered bacteria in modern bioreactors:

they are actually produced by plants which are harvested and treated to yield the sought-after drug. It's cheaper. And there is more to come. Right now, the chemical prospectors are beginning to comb the world for useful, not hitherto dreamed-of plant chemicals: the US National Cancer Institute is screening 10,000 such substances for use in the war against Aids and cancer. It would, of course, help if the plants hung around long enough for us to examine them.

We should be so lucky. Almost everywhere in the world, one tenth of the known plants are at risk of extinction within the next few years. In 1989 the USA listed 680 which should be extinct by the year 2000. Those, of course were only the ones the botanists got around to listing. The number — and the proportion — of plants that may be lost in the clearing of the tropical rainforests defies accounting. Nor are the plants the only loss to medicine: the animals — the reptiles and insects and even mammals and birds and fish — of the wilderness that may, no, *must* perish with them as well (and for that matter even the fish and the creatures in the sea) all yield alkaloids and hormones and other chemicals useful to medicine.

Some of them we have learned to value. Blowfly larvae synthesize something we now use for deep wound treatments; a European beetle provided us with a drug for some treatments of the uro-genital system. Medicine has need of armadillos (they, like us, suffer from leprosy) and seaweeds and coral polyps and the dogfish and horseshoe crab: there are lessons still to be learned from cold-blooded insects and fish that swim in sub-zero water temperatures but whose blood does not freeze because they produce their own antifreeze chemicals; or for that matter from that little reef fish which, as occasion requires, such as when the only male of the harem dies, spontaneously conducts her own sex-change operation, and thereafter, so to speak, wears the trousers; or from that quite surprising collection of creatures which, like us, reproduce sexually, that is, with a partner of the opposite sex, but when necessary can produce fertile eggs without any help from the male of the species.

But even here, in the field of mobile creatures like ourselves that base their lives on plants, we come back to the plants as the key to the creation of potential drugs and healing agencies: many of the plant toxins were evolved as part of the

great war of survival, for between plant and insect there has
been a chemical arms race, an escalation of offensive, defensive, counter-offensive measures, of feinting strategies, of
forms of detente and even peaceful co-existence and straight-
forward alliance, since first plant and then predator insect
began. The consequence is that almost any plant seems to be
equipped with an arsenal of chemical possibilities and chemical signalling equipment, sometimes never brought into play
during a particular life cycle, but there, all the same. Some
possess the machinery for shutting down utterly in a long
drought, playing dead, as it were and surrendering their arms,
among the most spectacular of these being the river red gums
of Australia, whose fallen, rotting branches and hollow
residual stumps allow a huge proportion of desert life to
shelter within them from the sun, but which shake themselves back into life, sometimes after years of apparent death,
at the fall of rain. Some plants have developed the equivalent
of gastric juices, and the apparatus to trap insects, some to
follow the sun, some to send signals of assault from pests to
other plants, some to come to amiable terms with a lesser
predator.

Wherever we look there are lessons to be learned from
living things; lessons to be enjoyed for ethical or aesthetic or
intellectual reasons — and economic ones as well. Not long
ago, scientists began looking at a tree common in India, called
locally the neem but known to botanists as *Azadirachta
indica* which also flourishes in Africa, and the leaves of which
were said by village wiseacres to have some health giving
properties, and which would protect crops against insect
pests. They isolated from the seeds of the tree an insecticide
which was effective at concentrations of a few parts per
million. They found that it disposed of at least 40 different
insect pests and yet was not toxic to man or to small mammals: it was not toxic because it operated in at least four ways,
picking off insects at various stages in their lives. In the first
place, it suppressed appetite; in the second it disrupted
growth; in the third it stopped the metamorphosis of a larva
into an insect and finally it worked on the insects themselves
by preventing them from mating.

This is a sophistication beyond the wildest fantasies of a
mad germ warfare scientist, a single chemical weapon that
inhibits the desire to eat, the desire to make love and engender

children, the mechanisms of growth and the triggers of adulthood, and still leaves bodies fit for a carnivore to eat; and yet at the same time offers no harm to the natural allies, in this case the bees and other insects that are necessary for pollination. It has been synthesized, that is made in the laboratory, from scratch, and tested on the Egyptian cotton leafworm: it works. It may soon have a role to play in Latin America, where 20 million people suffer from Chagas' disease. This tragic affliction is the work of a little parasite which lives and reproduces inside nerve and muscle cells, including those of the heart, leaving its victim too exhausted to work. It is carried by a bloodsucking bug which infests poor housing, and passed to humans through sore and punctured areas of skin. Azadirachtin does not kill the parasite. Nor does it kill the bug which spreads the parasite. But bugs which have been fed on blood laced with azadirachtin are, even 20 days later, nevertheless free of the parasite, because in some way the chemical — produced by a tree in India, remember — upsets the cycle of relationships on which bug and its parasite depend. It is one demonstration of the way in which one plant defence system can be applied to another set of human biological problems altogether.

There is a classic example of a plant defence with other purposes in the biology of the tree *Heavea brasiliensis*, a tree which has to use the brute force approach to survival. It grows on poor soil, away from the light, in the shade of other species: since it doesn't have the advantages of taller species, it needs an insect defence system that is secure. So any leafcutting insect that chews its way in or termite or weevil that bores into the bark soon finds its mouth and lungs filled with a viscous, white, sticky liquid that congeals swiftly: although this liquid kills mechanically, it also contains insecticides that wipe out those that survive from choking. There are, of course predators that find their way round such a defence. There is a caterpillar that has learned to cut off the channels in the leaves that carry this fluid before eating the edge of the leaf. There is another predator that actually attacks the tree for the fluid itself: by scoring the trunk and gathering the white secretion for its own sake. The secretion, of course, is latex, the predator is human, and the plant is better known as the rubber tree. For a material that changed the course of human history, we have to thank a family of plants that was forced to devise a means of survival on poor soils, in bad light, under the shade of bigger trees, and under continuous assault from thousands of potential predators.

We have barely begun to look at the other uses of plants in the Amazon. In one study conducted with the Chacobo Indians of Bolivia, a scientist pegged out the usual hectare and began identifying each tree, shrub, herb and vine. On that hectare there were 91 species of tree: 75 of those species were used by the Indians for some purpose or other, often medicinal, and those 75 species made up 95 per cent of that hectare. Many of these plants had more than one use. One, the guanara plant, has a high caffeine content and is used as a stimulant: it can also suppress hunger pains. The Brazilians make their own cola from it, and it grows well, like the rubber plant, in poor soils. The pataua palm is used by the Indians for food, fibre, building materials, medicine, weapons and even toys. Its fruit, when steamed and macerated, yields an oil with physical and chemical properties identical to the olive, and the pulp that is left has almost as much protein as milk and considerably more than soybean.

Another search unearthed five wild relatives of maize. Only two of them turned out to be food plants. One of them was a lethal poison. One was used to induce hallucination, the third to trigger abortion. The babassu palm, which also does well in poor land, is already in commercial use. A stand of 500 trees can produce 125 barrels of oil a year; and at the same time feed cake for animals, fertilizers, flour, high quality charcoal, methyl alcohol, tar and acetic acid. But the wild plants of the tropics, or for that matter the plants from almost everywhere on the globe are not important just as medicines to protect our bodies or insecticides to shield our crops or raw materials for our industries: they are important, in two ways, for our future food supplies.

If we are what we eat, then we aren't much. There are at least 75,000 edible plants growing on the surface of the globe. We have in all our history used about 3000 of them. Of these, only 150 have been cultivated on any scale. Of those 150, about 20 at the most provide more than 90 per cent of our food supplies. Rice, wheat, potatoes, soybeans, maize, millet, barley, oats, cassava, peanuts, tomatoes, rye, onions and variants of the cabbage family and some fruits — it depends where you are — provide the staple of the world. But the number of mouths that must be fed is on the increase: the chances are that we will develop sources of energy, protein and vitamins from other species, plants that we have neglected, or which are cultivated by local or tribal peoples, and the merits of which we have overlooked. Some of these merits are remarkable.

In New Guinea, there is the winged bean. The roots are edible and rich in protein. So are the seeds. You can also eat the stem, the leaves and flowers. You can make a kind of coffee from its juice. It can grow four metres or more in a few weeks. Crops of the winged bean are now being grown in 80 countries. The old Aztec empire once cultivated amaranth: it was as important to them as corn. Unlike corn, it can survive in dry lands, and it can flourish in both the tropics and the temperate zones. More than 12 per cent of the grain is protein, and a high percentage of that is a key amino acid rare in other grain crops, which makes its flour a useful addition to breads made from corn or wheat. The young leaves are edible, they taste like spinach. The Indians of western Mexico harvest a marine plant called eelgrass, the seeds of which they grind into flour: bread, so to speak, from the waters.

There are more than 1000 leafy plants in the tropical forests each of which contains at least as much protein as legumes and fruits, more than twice as much iron and 10 times as much carotene: they are in many cases also very rich in vitamin C and vitamin A. Then there are the gourds: the buffalo gourd in arid lands, for instance, can put down its tubers a good five metres in search of groundwater, and its fruits and their seeds are a source of oil, protein and starch. The wax gourd of Asia can yield three or four crops a year, and a full grown gourd may weigh 35 kilograms. The flesh is crisp and juicy, and because of the impermeable waxy coating, the gourd can keep, in store, unopened, and without benefit of refrigeration, for up to a year.

The peach palm from Amazona, when cultivated, produces more protein and more carbohydrate than the same area of maize: its fruit is also rich in carotene, niacin, vitamin C and fat. When boiled, it is reported to taste of something between a chestnut and a potato. Talking of which, there has been renewed interest in old varieties of this familiar plant: a recent examination revealed that one of the original Andean varieties had twice the nutritive value and much more flavour and better keeping qualities than the supermarket varieties now sold in Britain and America: being smaller, and more knobbly, it didn't of course, look as marketable.

But none of these new or revisited plants is likely to stop us relying on the wheat, rice, maize, potatoes or cassava that we already have as the staves of life, or replace as our major sources of protein the legumes: soybeans, chickpeas, lentils, haricot beans and so on. We rely on most of them because we have always done so: early peoples (maize and the potato are not new, they are only new to Europe) developed them for agriculture, and have gone on developing them. We know them well, we have built cuisines on them, and also whole worldwide economic structures. They are our history, growing at our feet, or at an arm's height above our heads, defining our landscape. You cannot write about France without mentioning the grape, or the Mediterranean without the fig or the olive, or explain the history of Ireland without word of the potato, or China or India without rice, ginger or the pepper, or the Pacific without the coconut or the yam. We cannot now imagine the sauces of Italy before the arrival of the tomato.

Invested in our crops are whole millennia of tradition, experience and experiment. Nutritionists, faddists, botanists and environmentalists may preach and promote new harvests: in many cases successfully. But the old ones are not going to go away simply because the colour magazines start praising the carob bean or the yeheb nut or (my favourite) the tree tomato, now diffidently marketed in Britain as the tamarillo. But our staple crops may not be here to stay. They too, may be at risk, ultimately, and simply because of the wholesale destruction of wild plant species, not just in the tropical forests, but everywhere.

CHAPTER 18

A stalk on the wild side

Little more than a decade ago, a Mexican botany student from the University of Guadalajara spotted something on a hillside that was, quite literally, unique, and utterly unexpected, something that botanists thought did not or could not happen. He discovered a plant known as teosinte, normally an annual, one of the wild relatives of maize, Indian corn, or *Zea mays*. There was a difference. Unlike other teosintes closely related to maize, this one was a perennial. That is, instead of growing a new plant from the seed of the year before, it stood and flowered and bore fruit from the same rooted stem, year in, year out. Like maize, and unlike other perennial teosintes, it had 20 chromosomes, which meant that it could be crossed with cultivated maize. It existed, very precariously, nowhere else but on that hillside in Mexico, and when the student stumbled upon it, the hillside was already being cleared and grazed.

The commercial significance, or for that matter the significance of such a find to the world's food supplies, cannot be overstated. Corn, or maize, is an old crop. The ancient civilizations of Central and Latin America were founded upon it: the plant was imported from the new world to the old along with tobacco, the potato and the tomato. It has become a staple of the Americas and of Africa, and it is cultivated almost everywhere in the world. It provides about one fourth of all the world's cereal grains and (because it is also a cattle food) one sixth of all the world's food. Derivatives of it are used to size textiles and paper, for cooking oil, for brewing, and in the making of penicillin, aspirin, chewing gum, tyre moulding: a thousand ways in a thousand forms of manufacture.

The history of the twentieth century may have turned on

it, in the strangest way. During the First World War, when the science of microbiology had not truly begun, a First Lord of the British Admiralty calmly asked a young, Russian-born biochemist working for Nobel in Manchester to find a way of making 30,000 tons of acetone, a solvent needed for the manufacture of cordite for naval weapons. The supply of imported acetone was limited, because German U-boats had begun to damage shipping. After frantic effort, the process was achieved: the scientist managed to find a brewing process that yielded acetone from, of all things, maize, and the British Navy's munitions were once again secured. Out of that process, the scientist established a relationship with the Prime Minister, Lloyd George, and in particular another Cabinet minister. The First Lord was Winston Churchill. The Cabinet minister was Arthur Balfour. The scientist was Chaim Weizmann. The reward was a formal declaration that changed the course of history and led to the foundation of Israel, with Weizmann as its first president.

But history could turn again on maize, and on its wild relatives, and on the unusual perennial teosinte on one ravaged hillside in Mexico. The USA provides 70 per cent of all the world's corn, and of that American harvest, more than 70 per cent is descended from just six parent lines. That is, it is inbred, it is refined, it has a narrow genetic base. It is a prime example of the Green Revolution, the new and remarkable range of skills, understanding and techniques in plant breeding which in the 1960s and 70s, changed the world's food supplies from deficit to slight surplus, the same revolution that turned India, poor, inefficient, sometimes ramshackle and often starving India, for a while into a net exporter of food. It was a revolution that took plants, scientifically selected them for their most heavily fruiting qualities, adapted them to specific climates, designed them to make short stalks so as to reduce the energy of growth and invest it in seed, and be able better to bear the weight of its own productivity, bred in resistance to many of the blights, rusts, mildews and weevils, encouraged them to endure droughts in one region, frosts in another, or flooding in a third, and still go on providing food for the world.

But by investing the future of the world's food supplies in super-efficient 'miracle' rice, or wheat, or corn, all of them based on carefully chosen selected strains, selected for this quality or that, the plant scientists invented a kind of monster. The monster is an upside down pyramid. It is at risk. It is not inherently stable. It can be toppled. Just as the plants of the tropical forest, or for that matter the desert, are in a kind of continuous biological war with pests, or climates, or soils, or mildews, so are farm crops. Crops lose their resistance to afflictions, or rather, the afflictions learn how to overcome the resistance bred into crops. Every new barley variety on the British market has lost its resistance to mildew within two or three years. The very use of 'monocultures' — one variety, sprayed with one fungicide, everywhere the eye can see — means that mutant mildews, subtle changes in pest attacks, arise swiftly. New pest attacks arise all the time. One day a pest may arise that could strike down 70 per cent of the entire US corn crop, or another that could lay waste the rice paddies of south Asia, or devastate the wheatfields of Europe or Australia.

This has happened in the past. The vineyards of France were devastated in the last century by phylloxera; the civilization of the Mayas in Mexico, who practised an intense monoculture, may have been weakened to the point of collapse in about 900 AD by the maize mosaic virus. In 1984, the farmers of Florida had to burn 6.5 million orange trees to halt the spread of a new strain of citrus canker; in the 1920s a similar strain forced the destruction of more than 20 million. In 1979 a blue mould cost the US tobacco farmers $240 million; the following year the same mould destroyed 90 per cent of Cuba's tobacco crop, shattered a precarious economy and converted the Havana cigar from an expensive luxury into an almost impossibly expensive one. And then there is the potato. The potato has been farmed in the Andes for at least 8000 years but the varieties in Europe in the 18th century were based on just two samples, one brought to Spain in 1570; another to England in 1590, and the entire population of Ireland were living on potatoes multiplied from a few clones of those. When late blight disease swept through the whole of Ireland, there was widespread starvation, profound political unrest and more than one quarter of the population emigrated.

No such thing happened in the Andes. The harvest there is not based on monoculture, but on hundreds of 'landraces' — varieties selected at different times and for different needs and purposes, and to overcome different stresses — and those landraces themselves were developed from different species of the original family of the wild potato. In food supplies, we buy inefficiency with variety, but we also buy insurance. In a field of corn, farmed by a peasant hundreds of years ago, or in some cases only dozens, there would be tall, thin plants and short fat ones, sturdy stalks that withstood the winds and slender ones on which, in a good summer, would be heavy and sagging with grain, some that went down with mildew or rust, and some that seemed to be victims of caterpillars and some that weren't; and in the next valley, or further up the hillside, the untidiness would be the same, but the mixture would be different, because the landraces were bred unscientifically, sifted to a certain extent as blighted stalks were handweeded, selected ultimately only for their capacity to survive and bear fruit. And around them, and indeed among them, if the crops were being grown in the country native to that crop — wheat in Greece or Turkey, corn in America — would be the weeds, which might not be just weeds, but in fact the wild progenitors themselves, the original wheats, the original maizes, completely unrecognizable as crops, but still closely related. It would look untidy.

It would not be economic by today's standards. It would not always provide enough food for the farmer's family and the nearby townspeople; it would, like all crops, be at risk from late frost, rain at harvest, storms, droughts. But consider the other side of the equation. Within that field would be the entire history of wheat or corn in that valley: the entire library, as it were, not just of the finest examples but of all the examples that had ever been. Farmers could not read that library, although they could make sense of some of it.

But they were its guardians, none the less. In such a field, genes are not lost. The genes are those tiny bits of biological information swapped in sexual reproduction, and passed from generation to generation. The importance to the plant breeders of having a huge pool of genes at their disposal was not much appreciated in earlier times: they did not have the wherewithal to recognize all of them, nor the technology to isolate them. We do have that technology now. Which is why, when the young student stumbled upon what is now known as *Zea diploperennis*, growing on a Mexican hillside, his discovery was hailed as the botanical find of the century. He had found something with the same number of chromosomes as *Zea mays*; he had found the potential for a perennial maize, he had found a plant which contained within it immunity to three major viruses which cost farmers millions every year; he had found a genetic structure which protects against the worms that infect stalk, root and ear, and offers tolerance in waterlogged soil. And he had found it growing on an area of only three acres, an area in which trees were being felled and cattle grazed.

But such a find, good news though it be for maize farmers, only throws into ever sharper relief the steady, continuing disappearance of all the old landraces, and the wild progenitors, of all the world's crop plants, both the ones we use now, and the ones we might turn to in the future. For they, too are being eradicated as surely as, and much more swiftly than, the tropical forests. At least 95 per cent of all the ancient wheats of Greece have disappeared in the last 40 years. Botanists have been combing Turkey for wheat's progenitors. They found some, and in those progenitors they found the genes for resistance to rust, an affliction of modern wheat: they found them just in time to limit an epidemic of stripe rust which threatened the whole US crop. There have been other close-run things. Just one small colony of wild rice in central India provided the only known genes for resistance to grassy stunt virus: the resistance was bred into a cultivar known as IR36, the rice that for a while at least fed the world.

But these are only a few instances of luck and success. The story of the wild crop plants is the story of wild species everywhere. They are at risk of extinction. Some are at risk because of war: Ethiopia, an unhappy land stricken by famine and bloodshed, happens to be one of half a dozen terrains rich in many of the wild species from which our crops have descended. But most of them perish as swamps are drained, grasslands ploughed or grazed, as new towns are built, as forests are felled and burned, as valleys are flooded for lakes, as industrial pollutants poison the waterways, as farmlands are sprayed with pesticides and weeds destroyed. For of course, to the farmer, wild wheat is a weed. And here too is an irony. Some plants show a remarkable ability to crossbreed with cultivated species. Most of them, of course, result in failures, or even more virulent weeds. But both rye and oats, those cereals which flourish where wheat or corn cannot, developed by accident from weeds in barley and wheat fields in the

Middle East and in Europe. Amaranth, beans, maize and squashes may have developed as weeds in fields planted for cassava and sweet potato. The wheat that provides our daily bread is a cross between emmer wheat and a wild grass.

The bitterest irony within the steady improvement of our crops, and the steady taming of the landscape all over the world within this century, is that we may be about to slam the door on our past, to shut out forever the possibility of the contamination of our crop plants, not from infection, canker blight, rust, mildew or insect attack, for they will always be with us, but from the weeds that may help them, those wild plants which, hardy, pushy, stubborn and vigorous, contain within them all the genetic material selected for survival: the very things that make them grow like weeds.

The other irony is that we have begun to understand this at just the point where we are about to eliminate between a quarter and a half of all the world's species. There have, over the past two decades, been efforts by plant breeders and botanists to establish 'gene banks' — frozen repositories of seed. But these, too, are not secure: an unnoticed power failure might wipe out an entire library of refrigerated material. This has been such a worry that two international agencies are planning to build a store in a disused mineshaft on Spitzbergen inside the Arctic Circle: here (for the time being at least, until the greenhouse effect starts disturbing the climatic pattern) the permafrost keeps the temperature below $-3.5\,^{\circ}$C. It would, in effect, be run like a Swiss bank: in which every country has access to its own 'safe deposit box' holding seeds of its own staple crops and economically important plants. At those temperatures, barley seeds should keep for 300 years; apple for 180; elm for only about 14 years.

But just because some seeds keep better than others, or need sharper temperatures, and just because we don't really know how long seeds can remain dormant in low temperatures (barley seed, of course, has never actually been kept in a refrigerator for 300 years) there have to be other solutions. These include botanical gardens, parks and orchards where the wild and landrace varieties and old cultivars can be conserved and allowed to reproduce. These are expensive to maintain, and it is difficult to persuade politicians that, say 3500 or even 350 varieties of English eating apple must be preserved forever when the supermarkets only sell three.

Once again we face a terrible irony. Food is the world's greatest industry; it is the underpinning of all our economies; without food, all our other efforts are purposeless. Without the assurance of food, all our civilizations would crumble within a week. For food, read plants: and here we are, our colonization of the entire globe already so thorough that people in expensive suits in Rome are proposing that a hole in the ground in the Arctic Circle should be the repository for the past and therefore the future of agriculture for the whole world. A hole in the ground on an island north of Norway and the Soviet Union that will hold the past and future of the coffee of Brazil, the millet and soybeans of China, the potatoes of Idaho, of Jersey, of the Andes, the maize of Mexico and the yams of the Pacific and the wheats of Europe. And other people in suits and white laboratory coats in the UK and the USA and Europe are pleading for funds to buy and maintain land to shelter and nurture what a few years ago were regarded as useless weeds. But, ultimately, neither frozen gene bank nor botanic garden is a solution. A wild plant in a botanical garden is much like a tiger in a zoo: if it does reproduce and flourish in a zoo, then a tiger ceases to be — in a number of important ways — what we understand to be a tiger.

A plant in the wild is a survivor of thousands of years of random assault from predators, drought, flooding, fungal and viral attack, and its genes contain within them the chemical formulae for a thousand resistances to a thousand different attacks. That, ultimately, may be why we need the wilderness, and all the creatures in it. They are survivors. We will always have something to learn from them. Without them, we in the comfortable West may instead learn the meaning of famine, when whole prairies of wheat, all of it descended from a few strains, and all of it at risk in the same way and at the same time, fall prey to a new assault, one we had not been expecting. But then much of the rest of the world already knows a great deal about hunger, and hunger's ever present attendant, disease.

CHAPTER 19

The kingdom of sand

Every year an area of land the size of the Republic of Ireland begins to turn into a desert. Put it another way. Every two years an area of land the size of Cuba becomes a desert. Or another way: in three years, an area of land the size of Cambodia has become a desert. In four years, an area the size of the United Kingdom has become sterile. It would take about eight years for a country the size of Iraq to become a desert but, of course, part of Iraq began as a desert and much of the rest of it has since become a desert anyway, not despite but *because of* human attempts to make the land more productive.

If you count as a desert either terrain or water by which humans once might have successfully grown food and on which we cannot now do so, the figures above are conservative. Strictly speaking, a car park is a desert: so is a skyscraper city; so are the urban shantytowns that swell round the capitals of the poorer nations; so are the gardens around the industrial zones of say the Rhondda Valley, parts of Poland, and parts of Japan where the outpouring of toxic metals from the smelters of industry has been so steady and so furious that the fine dust of the metals has been taken up by the plants, making them unfit for human consumption; so too for many decades was the island of Gruinard off the coast of Scotland where British scientists, in the interests of peace and defence, tested the bacteriological warfare agent — in this case, also a natural affliction of farm livestock — anthrax. So too is the farm and forest land around Chernobyl in the Ukraine, the scene (at the time of writing) of the world's worst nuclear accident, dusted with prodigious quantities of fallout in the form of caesium 137; so are some of the sites where nuclear weapons tests have been conducted; so, too are the rivers, estuaries and offshore banks that once bore spawning populations of fish but which have now been choked or poisoned by industrial effluent, sewage or household detergent from cities and from heavy industry.

That vast body of water, the Black Sea, might be heading swiftly towards becoming a sterile zone. Ninety per cent of the seawater is now contaminated with hydrogen sulphide: there is so much of this essentially toxic gas saturated in the water that a small earthquake might send colossal quantities belching from the surface. At a guess, there might be 7600 million tons of ammonia, ethane, methane and hydrogen sulphide dissolved in the lower waters of the Black Sea, all of it effluent from factories on its shores; so much that it might be possible to found other industries simply to reclaim it, and create new wealth out of the excrement of the old, and at the same time begin to restore the Black Sea to its potential as a producer of food. But all that is by the way. These things do not count in the statistics of the making of deserts.

The classical definition of the desert will do for a start: the desert of the television screens; the deserts from which the sand debouches, spilling inexorably in dune formation, marching south from the Sahara over grazing land at the rate of a kilometre a year, choking dry riverbeds where the last moisture has lingered below the baked surface, burying oases and smothering the hardy acacias and thorns on which some animals might have browsed; the desert that obliterates, if not life in its entirety (for life is ultimately persistent, stubborn, cunning) then any chance of human survival. And then beyond that the classical arid lands, the dry lands, the sub-humid tropics, the soils which at first look bleak and fruitless, but which are not, and from which 850 million people make their livelihoods; these too are dessicating; on them the tribespeople and the farmers and their families huddled in small townships and by shrinking rivers watch their meagre millet crops straggle and — with greater frequency — fail, their cattle and goats wander further and further for forage, becoming weaker and then collapsing and dying; that land on which the women and children walk an ever greater number of miles each day to gather ever fewer sticks of firewood to cook; that land on which even the animal dung which should be returning nourishment to the weakening soils is scooped up and dried and burned to cook the dwindling meals: that land from which columns of refugees every season leave their homes and their hopes and trudge towards (but many do not arrive, falling and dying on the way) the refugee camps already full of sticklike bodies which can do little more than cry a little, and hold out fragile arms for a dole of food, and nurse dying babies, and wait for the rain, or death from malnutrition, or death from infectious disease, such as cholera.

And beyond those arid soils, the lands that were once successful farms, and on which the rains had, for a year or two, failed; land with roads, and towns, and reservoirs, and woodland, and fields, and large pastures, and plantations, land still marked by the trappings of human economic success, and already dusted with that trademark of failure, literally, dust. Enormous tracts of eastern Africa north and south of the Equator, a land marked with vast lakes and mighty rivers on the map, have from the air already the look of a desert. The plains and savannahs and tablelands are baked reddish brown, and the bands of vegetation near the water supplies are not green but that sun-bleached, greyish colour of plants whose roots are touching naked rock and wrung-out sandstone. From a height, it looks like a desert. As the plane descends, it doesn't look like a desert because you can barely see at all: at a hundred or so metres the aircraft descends through swirls of fine reddish dust that almost obscures vision, and then on the ground the picture changes again, but not much. There is vegetation, but it too is dusty, and yellowing. The people move slowly, sometimes cheerfully, sometimes with a fatal listlessness that marks the beginnings of despair but the signs of the true hopelessness of the desert are not so easy to read.

You see them sometimes in the trucks laden with cornmeal coming not from the villages to the towns but the other way, from the towns to the villages. In the cities you can see the auguries of hunger and impoverishment on the office nameplates: Oxfam, Save The Children, CAFOD, Action Aid, FAO, WHO, Unicef, Red Cross; you can see them in the knots of people walking into town carrying pathetic bundles of dried sticks to sell for firewood; or in the shaded wayside stalls by which people sit patiently, with only a single mango or a few bananas to sell; you can see them in the growth of the shantytowns on the outskirts of modern cities, and in the groups of women and children standing by standpipes of water, standpipes used at so many points along their length that at the end the water comes out only in droplets. Welcome to the kingdom of hunger and sickness and hopelessness, the biggest single community of all, a community that could one day extend to embrace a number not much smaller than the population of China, a community that circles the world in a huge band, a community that extends its borders by a few miles every day, a community whose inhabitants range from the dull-eyed and the emaciated, weeping with dry eyes over the tiny bundles that were once children, to the still sturdy men asleep of a Monday morning in the sun in the doors of their breeze-block and corrugated iron homes, asleep because there is no soil to till, because they have been forced to leave it, nor paid labour to be gained in their new homes: they know, because they have tried, and tried again, and gone on trying and eventually found that it is better to sleep, to blot out for a few hours the emptiness that stretches ahead.

Welcome to a community that is constantly changing; where the aid agencies are at work on land rescue schemes, or where the rains have returned, and the harvest been big enough to provide a sufficiency, or even better, a surplus that can be sold; or from pockets of which young men with some education have been able to migrate, and earn money to send home to buy food and pay debts, or buy visas and aeroplane tickets to a new life; welcome to a community from which every year a few hundred thousand escape, and which a few million join, either temporarily as rains or harvests fail in India or Africa or China or Brazil, or for a lifetime, a lifetime shortened either by hunger or illness or war. Welcome to a community in which the older men and women sit dully, conserving their strength and their younger, angrier brothers and sisters and their older sons and daughters turn to violence, robbery, prostitution, drug-smuggling, ivory poaching, murder and terrorism not because it is the easier way but because in a shantytown, or on the streets of a big city, it sometimes looks like the only way.

Welcome also to people who — in those dusty villages and tribal settlements with precarious water supplies, no telephone, where letters arrive haphazardly or not at all, where there is almost no wheeled transport and often no beasts of burden, no electricity, no medical services, with schools but with no writing materials or books to read, and with no sure nourishment but cornmeal and fruit and a little hope — can still conduct themselves with dignity, good humour and patience, and can welcome European visitors with dignity, courtesy, good humour and an intelligent, perceiving friendliness that, unaccountably, can match any they may find in their own homes, and cause them to think (they would not be the first to think this) that people who have too little may not be less happy or human or less innately decent than those who have too much — this should, of course, have been obvious all along. Welcome to tomorrow's desert.

Deserts exist because geography dictates so. They, like the forests, are part of the engine of climate, and like the forests they are also the consequence of climate. But they are not fixed. The study of the rocks teaches that, aeons ago, there were deserts where there is now fertile land and rich soils now lie on top of sandstones in which you can still detect the patterns left by shifting desert dunes. But now the spread of the deserts is because of the abuse of the soil. Unfortunately — because the spread of the deserts is the most complex single problem with which we now have to deal — there is more than one way to make a desert.

Deserts happen because we chop down trees. It isn't as simple as that and it isn't the only reason, but it is ultimately so. Trees shade ground and conserve soil moisture; they hold shifting soils stable and they reduce the loss of mineral nutrients every time it rains and their fruit and lower leaves provide forage for grazing animals in a manner that doesn't disturb or loosen the soil, and the loss of foliage is compensated for by the return of dung to the ground beneath them. In a clear-felled area of New Hampshire (never likely to become a desert, in the classical sense) scientists measured the mineral loss from the soil and compared it with that from wooded land and found that minerals leached from the soil three times faster from the cleared land. Also, in hot, sunny, arid climates, trees lower the albedo and make the return of the rains just that little bit more likely, and their loss raises the albedo and makes it just that bit less likely. Drought and famine — the indicators of a desert in the making — happen even in lands we think of as well watered: in the last few decades millions have been stricken by occasions of drought and, because of drought, famine, in north-eastern Brazil, or in Kalimantan in Indonesia, landscapes we still think of as covered by rainforest, the torrential rains once as regular as clockwork, but no longer.

Deserts happen because the poorest people change the way they use the land: in the driest grazing lands, they settle and plough instead of following their herds in search of forage across the plains, and then the soil around them collapses under the combined stress of overgrazing and cropping; and the gathering of wood for cooking fuel strips the soil of even those stunted trees and shrubs that might hold it stable to recover in later years. Or they increase the density of grazing, so as to denude dry grasslands in a year or two, or they use cattle where goats would be better, or goats where only a camel could thrive. Under pressure of grazing, the soil structure in which grass normally survives begins the process of collapse; after that, the topsoil begins to move in the rain, bakes dry and powdery in the long months of sunshine that follow, and begins to blow away.

Topsoil — our greatest asset — is a fragile thing: a barely-understood microcosm of minerals and substrate and humus and microbes. It is the slow creation of the millennia, a thin skin of survival's provender, made by the action of microbes and compost and rain and sunshine and gentle flooding and frost and root-action, by the burrowing of worms and beatles, by the excrement of animals and the annual benison of leaf-fall and dying grasses, until it lies almost everywhere in the world — except the deserts and the ice-caps and the implacable rock-faces of the highest mountains — to the thickness of a finger or even metres deep, to provide the raw material of survival. It is the biosphere's bottom line.

But it blows away in the wind and it runs away with flooding and in ploughed land, or over-grazed land, it blows or seeps away far faster than it can be replaced; sometimes by the ton per acre, even in temperate lands; and its loss, even where the topsoil is not at risk, exacts a price for humanity. There is a strict relationship between crop yield and topsoil depth. The deeper the topsoil the better the yield. At a certain level of shallowness, yields fall away dramatically, however gentle and steady the rain. And where there are long periods without rain, the land degrades still further, and more topsoil blows away. It is happening most dramatically in Africa, but it is happening in China, Central Asia and parts of the Indian subcontinent: it is happening, surprisingly, in the American west, where to meet an upswing of consumer interest in 'free-range' healthy and environmentally friendly beef the ranchers are overstocking the federal lands of 11 states, stripping the ranges of herbage and grass, and trampling the soil. (A bullock's tread has a pressure of 1.7 kilograms per square centimetre.) In some places half the original topsoil, the topsoil that once supported — and even increased while supporting them — bison, antelope, wild horse, Sioux and Apache, bighorn sheep and elk, and a carpet of wildflowers, has already gone. And with it has gone the watertable: creeks and springs are now dry. About 10 per cent of US land, all of it in the west, is now turning into desert because of overgrazing.

There are other reasons, all ironic. One of them is simply the flight from the land. With the oil boom in the Saudi Arabian peninsula, the ancient peasant struggle to win a living from the land collapsed; there was no need for it; there was work and wealth in the towns. So the old hedged and watered terraces that won crops from the hillsides began to collapse, and the topsoil nurtured by hard labour for generations simply slid downhill and blew away. Another is irrigation, the tool that humans first devised to make the dry lands bloom. One has to be careful with the use of water. If the land is not properly drained, or if dry soils are watered too copiously, they waterlog, and the water becomes saturated with salts dissolved from the deep soil. As the water evaporates more rises to the surface and evaporates again, leaving the topsoil with extra salt in it. Or the water itself is slightly saline and the constant cycle of irrigation and evaporation begins to leave a crust of salt behind it. Some crops can cope with slight, saline soils, but there is a limit to how much salt any of them can cope with.

Salt is a problem even where irrigation isn't needed. On coastal zones, the clearing of trees itself lowers the watertable, which means that the local people need more and deeper wells, which lowers the water table still further. But coastal water tables float on a layer of salt water: sooner or later this too begins to penetrate the land. Vast projects in Africa, Iraq and India and Pakistan and the Soviet Union, have foundered not because of too little water, but because of bad management of too much water, which left the land poorer and more sterile than it was before, some of it irretrievably so, and salt flats have formed even on the moist and fertile coasts of tropical west Africa.

At bottom, 35 per cent of the world's land surface is at risk of turning into desert. This land is home to 850 million people. There are 45 million square kilometres of drylands. Of these, an area which is about as big as North and South America combined is already at risk of turning into desert. Each year a large chunk of this actually becomes desert: that is, useless land from which the only harvest is dust and despair.

It doesn't have to be like that. In the first place, life itself is stubborn, and springs forth in the harshest deserts and arid lands after a fall of rain. In the second place, Nature (remember, this is simple shorthand for a complex of evolutionary pressures) has provided a number of potential solutions, and as in the rainforests, a number of plants of forage and medicinal value. One of these is used in Ethiopia to control schistosomiasis, a wretched and crippling disease spread by the snails that infest the water courses: its berries exude a molluscicide. (Once scientists started looking, they found natural plant molluscicides almost everywhere in the tropical world.)

We have used some of the desert plants for centuries, and this use is now being extended by science. There are grasses that show a remarkable ability to hold fast in shifting soils, and in doing so, hold the soils. There are salt tolerant shrubs that can feed and fatten camels, and camels could be the salvation of pastoral Africa, as they were once for the Arabian bedouin and the Tuaregs of the Sahara. A female camel lactates for a year after giving birth, and produces 10 litres of milk a day when pasturage is good; in the bad season it will produce as much milk as a cow will give in the same regions at the best of times. Their feet don't damage the soil surface and camels roam widely, destroying less vegetation and browsing on trees. To produce a litre of milk for a human a cow must consume nine kilos of dry matter; the much more efficient camel, two kilos. They can, of course, survive long periods without water.

And then there are the crop shrubs. Euphorbia and jojoba

are already used by industry (euphorbia produces an oil substitute and jojoba produces a lubricant almost identical to that from the sperm whale). They are tolerant of salt and drought, and they can provide a cash income for desert farmers. There are techniques for using brackish water to cultivate celery, aubergine, tomatoes, asparagus and melons, all of which are relatively salt-tolerant. (In fact slightly salty water actually triggers an increase of sugar production in grapes and tomatoes.) Both the beet and the date palm are descended from salt tolerant ancestor plants. A very edible New Zealand spinach tolerates both hot weather and highly saline water. Altogether there are more than 1700 known halophytes — salt-loving plants — which can play a part in regenerating wasted land. Then there are the trees, the acacias, the date palms and the pines and eucalypts that seem to thrive on sand and survive long periods of drought. There are others, only just being examined.

Of these the desert-loving prosopis of Afghanistan, Oman and Iran must be the most extraordinary. It can survive as a bush a metre high or a tree ten metres high. It throws down a tap more than 30 metres into the desert sands in search of water, and can bear leaves and even flowers during a severe drought; it can survive fierce cutting back by humans; it is a legume, so that it actually adds nitrogen to the soil rather than takes it away; its leaves and pods provide food for the animals, and it has been found surviving in deserts even where the water table is 30 metres below the deepest taproot because it can absorb the night dews of the desert directly through the pores of its leaves. It can propagate by cloning and it seems to survive burial by dunes, simply by growing through the sand and forming a new canopy above them. It plays host to lichens and can be eaten by gazelles, goats and camels, and scientists of the Royal Geographical Society in Oman counted 28 bird species nesting in prosopis woodland, demonstrating that in the midst of apparent sterility, life goes on.

None of these things are solutions: they are simply further potential tools in what will have to be a concerted and intensified international effort, probably led by the United Nations Environment Programme, using public education, local co-operation and government persuasion at every level of every society, together with a mixture of traditional and scientifically-tested techniques, of both low and high technology, to halt the spread of the deserts and restore the lost land to some kind of economic use. Such a programme exists now, but it is neither concerted nor intense. Without a renewed effort, the sands will go on marching, and the skies of Africa and Asia will go on darkening with windblown dust and yet more millions will leave the land and begin the long walk to the towns and the aids camps.

There is already a new class of displaced person: displaced not by war or pogrom or political persecution but by environmental degradation, and this class already bears a label: environmental refugee. Their numbers are swelling. By March 1985 the drought in Africa had scorched and seared the land so severely that more than 10 million people in 24 countries had to leave their homes, and in some cities the population increased fourfold. This, too, will go on, and so too the appeals for emergency aid — ultimately useless aid in that it keeps the starving from death without restoring the land on which they depend — will go on, and become more strident, and more tragic, until we too become deaf to them, and start worrying about the high price of our own food, and the sudden shortages of some crops to which we had become accustomed. But by then wars will have started, as they used to start, over simple basic things like food, and land, and water.

CHAPTER 20

The dominion of death and life

In 1898 a man called Newton Harrieson put forward, in his journal *Electrical Age*, a new plan for world domination. He proposed a 25,000 mile, ten-strand electric cable, each strand 12 inches thick, right round the globe, powered by a current of a million amperes. At 10 million volts pressure, this current would be strong enough to shift the north and south magnetic poles, convert the globe into a vast electromagnet and change the tilt of the earth towards the sun (the sun, he thought, being mostly iron). This would permit the melting of the ice-caps and the conversion of the tundra into a watery garden, and would temper the heat of the tropics. 'Gone would be the yellow-fever of Cuba, orchards would take the place of the jungle' he rhapsodized. He also argued that warfare would disappear: it would have to cease because the nation in charge of the current could annihilate enemy nations at will: 'The United States at a touch of a button would transform them in turn to frozen wastes or torrid deserts . . . Ours would be the victory without a single loss of life.'

It was a great year for world domination: in 1898 Nikola Tesla proposed a guided missile — he called it a radio-controlled dirigible boat but it sounds just like a torpedo — that would abolish war by rendering coasts impregnable and the mightiest navies so much scrap iron. (These and other enchanting ideas are to be found in *A Victorian World of Science*, a study of early technological magazines written by the Cardiff scientist Alan Sutton in 1986.) But since then we have grown used to technical projects for world domination. By the early years of this century one small group of farsighted people had begun to argue that modern warfare would be unthinkable simply because the power of artillery and high-velocity small arms would impose an impossible cost. Another group had begun, more realistically, to foresee an age of Total War, in which citizens would be as involved as the military. Both were right, and the arms race that had already begun in Victorian times ended in the 1914–1918 war, also known as the Great War and, because of the scale of the slaughter, also the War To End All Wars.

It didn't, of course, do any such thing. During the Second World War between 1939 and 1945 more than 30 million people may have died and this too was followed by a number of smaller civil and military wars — at least 120 in all, in Asia, Africa, Central and South America and even Europe and in these conflicts another 20 million may have died. During the course of these at least two major technical developments emerged with the declared aim of outlawing war. One was the buildup of intercontinental ballistic missiles; the sum of the two arsenals being enough to obliterate all life on the globe, and their existence a tribute to the doctrine of Mutual Assured Destruction, or MAD, which argued that war would be impossible because there could be no victor. A second project, still half-heartedly pursued, is President Reagan's Strategic Defence Initiative, SDI, or Star Wars, which proposes a global umbrella of space-based laser weaponry designed to detect and shoot down enemy missiles before they can reach US territory: a project designed, like Newton Harrieson's to leave the USA as undisputed master of the globe; invulnerable to assault, but able to destroy enemies at will.

Like Newton Harrieson's plan, SDI in its overripe form is a rotten idea. Harrieson had done his sums wrong and it wouldn't work, and anyway even if it did no nation in its right mind would let the US complete the cable. SDI has met much the same objections, the most cogent being argued by the planetary scientist Carl Sagan who has also pointed out that even if it did work its management would have to be conducted by a computer so sophisticated that even its software would have to be written by a computer; and that under its own internal logic the system would have to be constructed so as, of its own initiative, to eliminate any surprise attack, begin reprisals, meet any second offensives with counterforce and so on. Therefore it would have to be a computer system that could declare the Third World War open before any humans in the world knew it had begun, and go on waging it long after all humans in the world were dead.

But the world-encircling cable, MAD, SDI and all the other ideas that have popped up in each lap, so to speak, of the arms race — such as chemical and biological warfare, the neutron bomb and so on — have turned out to be beside the point. World domination seems to have happened already anyway and we haven't even really noticed. For the last 10,000 years of our existence, humanity's chief struggle has not been against fellow human beings, but with Nature, with the wilderness, with the obduracy of the earth: hunger for most people has always been the real enemy. It has been this way in Europe for most of our history: that may be why so many of our old fairy stories — the magic tablecloth, the gingerbread house — centre around food. But if there has been a struggle with Nature, we may be on the verge of winning it. Our victory seems so complete that the larger creatures of the wild will only survive on our say-so. If the whale survives, it will be because we have taken a decision not to hunt it. If the elephant survives, or the tiger, it will be because we have made provision for it to survive: caged it in a reserve, and managed it as surely as we manage animals in zoos. If the tropical rainforests survive it will be because of conscious decisions by governments, by the World Bank, by industrialists. And the almost total victory has been achieved not by technical prowess in the cause of destruction, but by two forces for unequivocal good: love and medicine.

There are, it is sometimes argued, more people alive today than have lived during all previous human history. If this is true, it is because the growth of population has been exponential. Think of a number, double it, double that number and so on: 1, 2, 4, 8, 16, 32, 64 . . . The sum of the numbers from 1 to 32 total only 63: that is, the last number in the sequence is always greater than the sum of all previous numbers. Notice also that the sequence begins slowly and accelerates: that is there are four numbers below 10, but only three below 100; the next sequences goes 128, 256, 512 and then suddenly the 1000 is breached. This is exponential growth, constant doubling. For the first 10,000 years, growth of the human population was slow and probably fitful: Athens can have been no more than a small town by our standards; Imperial Rome a small city; the Mongol hordes and the tribes that rode with them, that swept all of Asia under Genghis

Khan, have been put at no more than 250,000. At intervals the population of various parts of the world was dramatically reduced by plagues such as the Black Death, at all times infant mortality was high, and the average life span was short.

By 1800, the beginning of the Industrial Revolution and the point at which we started to alter, at first imperceptibly, the machinery of the globe by burning more carbon-based fuels than the plants of the earth could absorb, the entire human population had reached probably one billion. Fifty years ago — less than a lifetime — the population of the globe had reached 2.5 billion. Today it has passed five billion. This has happened despite the staggering destruction of the wars of this century, despite the colossal waste of life in natural disasters such as floods, earthquakes and famines. It has happened in spite of the environmental insults — photo-chemical smog, acid rain, water pollution, the spread of substances feared to be carcinogens such as dioxins and PCBs, in spite of the toxins in the food supply, in spite of the rain of arsenic, lead and cadmium almost everywhere from industry. The world could hold six billion people by the end of the century; eight billion thereafter, perhaps 10 billion before the end of the next century.

This has happened, of course, because of the discovery of the merits of hygiene (in the seventeenth century people in Europe were inclined to think that filth drove out contagion) and the laws against contamination of food; the discovery of public health techniques like quarantine and the cordon sanitaire against plague, the development of clean water supplies and sewage disposal; the development of surer techniques in the treatment of disease; in the recognition almost everywhere of the importance of balanced diet and the growing understanding of nutrition; and in the develop-ment of vaccines and drugs and disinfectants. Smallpox has been eliminated altogether; by the year 2000 the World Health Organization plans to be immunizing every child in the world against six major diseases; it also plans to eliminate forever the wild polio virus.

The consequence of this series of decisions for good are visible in almost every Third World nation: a huge preponder-ance of healthy children and teenagers, many of them there-after destined to unemployment and a life of hunger and frustration. Even so, the figures are startling. Since 1960, the

World Health Organization, Unicef and other bodies have reduced infant mortality rates steadily. Then, of 1000 children born worldwide, 127 died in infancy. Now only 72 die. This means that every year an extra seven million children survive their first year of life. In 1960, some 69 children out of every 100 used to die between the ages of one and five. This figure is now 39. So every year an extra four million children survive to the age of five. In effect, 11 million deaths are now averted every year. These figures are likely to get even better.

There are signs, however, that the rate of population growth may be slowing. There is evidence that as infant mortality rates fall, so do the number of births. This too seems to be happening not only in those countries with welfare programmes, high employment, strong economies, access to birth control methods and universal literacy, but even in those nations which have none of those things. In 13 European countries, the number of births is already balanced by the number of deaths; some kind of stabilization is expected to occur eventually in those countries which are now growing fastest. According to a series of statistics unveiled before an informal meeting in 1988 before the chiefs of the World Bank, Unicef, WHO and other bodies such as Rotary International and the Rockefeller Foundation, something amazing could happen by the year 2000: possibly for the first time in human history there could begin to be a decline in the total number of babies born each year.

There are factors that make even these predictions complicated. One of them, of course is the disease Aids, that puzzling, protean virus which after years of dormancy suppresses the immune system and ends, as far as is known, always in prolonged sickness and then death. It was first identified only in the beginning of the 1980s, and already threatens the lives of between five and 10 million the world over. Its hold in the populations of America and Europe is chiefly among those homosexuals who are promiscuous, among drug abusers, among groups such as the haemophiliacs and others who absorbed the virus through blood transfusions, but that does *not* mean that its grip is only upon the minorities of society.

In Africa the spread of the disease is through the heterosexual population: babies are born with it, and in some cities of the continent surveys have revealed that between 10 and 20 per cent of pregnant mothers carry the virus. Any children they have will carry the virus, and will not survive to maturity. It is therefore possible that by the year 2000 Aids alone, in Africa, will wipe out all the gains in child mortality since 1960. On the other hand the disease may gradually be controlled by public education; by the lessons of other international research campaigns against pandemics, it may eventually prove to be possible to check it or cure it. There are other imponderables. There is the damage to the ozone layer: the extra ultraviolet radiation may suppress human immune systems and increase the spread of infectious diseases, but this is conjecture. Nobody knows. There is the buildup of greenhouse gases and the consequence of a warmer world. Dengue, yellow fever, malaria, leishmaniasis are all tropical diseases: their range may spread as the temperate climates become warmer.

Another imponderable, oddly enough, is military spending. There is an apparent correlation between the proportion of its gross national product that a nation spends on arms and child death rates. That means that the countries that spend proportionately more on guns, planes and tanks also have higher child mortality rates: which in turn suggest that birth rates will rise. But even the emergence of Aids and the huge international spending on arms has not seriously dented the projections of falling child death rates and therefore, at sometime in the next ten years, the possibility of an actual decline in global births.

However, this much looked-for fall does not mean that the population of the world will start to decline, because modern medicine and public health techniques between them have achieved something else: they have extended the potential life span of humanity everywhere. The average life expectancy of a human being is now 60 years; in the year 2000 it will be 65. Expectancies vary from country to country but the improvements continue even in the poorest. In 1960 there were 10 nations whose people were born with an average life of 35 years or less, now there are none. There are more than 40 nations with average life expectancies of 70 years or more. The gains continue, literally every day. Life expectancy in Asia has increased dramatically: for every day in which an Asian has survived, his or her life expectancy has increased by 10 hours. This is why, even though the overall number of births each year in the world may begin to fall, the population will still soar. Kenya now has 23 million people: its population is growing at 4 per cent per annum, which means that it will double every 17 years and thus it looks forward to 121 million before it can stabilize; Nigeria has 112 million; it could end up with 529 million; India with 812 million now will end up with nearly 1700 million before births and deaths are expected to balance each other.

Once again, we are faced with numbers that defeat the imagination: the population of China *now* is greater than the population of the world in 1800; the population of Mexico when it stabilizes will be greater than the population of the USA is now; the populations of India, Bangladesh and Pakistan, when these stabilize will be greater than the population of the whole world when these nations became separate geographic territories after the Second World War. Once again we have to think two things at once: we have to set the unalloyed good — the joy to a single family of a child that lives, and the security within it of parents who can expect a healthy life and a dignified old age — against the global or territorial bad. Not least of these will be the make-up of the population: the proportion of people who survive beyond 90 or 100 years will grow, and the biggest growth industry in many countries could become simply the care of the aged.

But there are more serious worries. Among these is the diminishing hope that must confront the survivors in such countries, for in all of them, the land available for growing food is shrinking, not simply because of the loss of soil and surface for building, but because of environmental degradation, and the forest and the grasslands and the fertile soils that once defined them as lands of hope are diminishing, perhaps forever, the last literally blowing away in the wind. Right now about 700 million people go to bed hungry every night. Soon there will be a billion people on the edge of starvation.

Where will they go? We know the answer to that already. They will move to the cities. By the year 2000 nearly half of the population of the globe will be living in cities. In the next 10 years another eight million people are expected to live in Mexico City, another five million in Calcutta, another six million in Bombay, another eight million in Sao Paulo, five million in Karachi, five million in Djakarta. Three quarters of all the people living in Latin America will be living in cities by the turn of the twentieth century. But these people — most of whom are alive today only because of the good things which have occurred as a result of science and civilization over the last century — who have turned to the towns for hope because of the blight of the land, are part of the threat of tomorrow.

If the rural areas cannot feed them, how can the towns? A growing city eats land just in its growth (in eight years between 1967 and 1975 the USA lost 2.5 million hectares of farmland to urban sprawl). It eats more than land: it eats the produce of the land. A city of a million consumes 625,000 tons of water; 2000 tons of food; 9500 tons of fuel. The fuel, the food and the water come from the rural areas. The same city produces 500,000 tons of waste water; 2000 tons of rubbish and 950 tons of air pollutants. All of that has to be disposed of in the rural areas.

Nor is famine the only thing to fear. None of the resources of the globe is limitless. In some nations, energy supplies — from oil reserves to fuelwood — are disappearing at an accelerating rate. There is already international alarm about water. There are, roughly, only about 9000 cubic kilometres of fresh water available at any time (in streams, creeks, springs, wells, lakes and reservoirs) for humans to drink, cook food, wash themselves, and use for agriculture and industry. Right now we are using about one third of this. In theory, this supply should be enough for 20 billion people. But 20 billion people doesn't sound such a large number, in a world heading towards a population of eight billion. And of course, the water is unevenly distributed. Some nations are already desperately short of water — and land on which they can grow crops. Others are already dangerously near the limits of their supply. The stores are disappearing before our eyes. In Central Asia, the Aral Sea is drying up. The level of Lake Baikal in Siberia is falling. Nations are already competing for the benison of the Nile, the Jordan, the Ganges and the Brahmaputra. And of course, what we don't use we pollute, with sewage, with industrial effluents, with fertilizers, with toxins.

And all this is happening in a world which spends a million dollars on military equipment every 35 seconds. Three weeks of global military spending could provide primary health care for every child in the world, and access to safe water; a mere six hours would be enough to stop five million children dying of diarrhoea; 10 hours of military spending would be enough to supply 80 million women who want birth control techniques with contraceptives; two and a half months of military spending might be enough to reshape agricultural production in the poorest nations, and build up once again the fabric of hope. There is, so far, no evidence that any such things are likely to happen. All the portents are that the violence will get worse. Those nations without resources may have to resort to force to acquire them, and those who can see their own resources dwindling are likely to defend them with force. There is, after all, not much point in having weapons unless you are prepared to use them. It might be that the advances of Lister and Pasteur and Jenner have all been for nothing, and the logic of world security has once again been invested in people with the mentality of Newton Harrieson, which would be a pity.

It would also be an irony. After all, Harrieson's programme for world domination involved the construction of a weapon which would alter the climate of the globe, and, as we have seen in the first half of this book, we may have brought that upon ourselves anyway. The difference is, of course, that were Harrieson's project to have worked according to his dream, it would have enabled us — or at least the controllers of the cable — to alter the climate at will, with (he thought) predictable consequences and beneficial results. The consequences of the human experiment of the last 200 years, and the massive industrial expansion still to come, are not at all predictable. Nor are they likely to be, on balance, beneficial. The only comforting sign is that there is almost certainly more concern about the global environment now than at any time in human history: indeed, until this century there has probably never been any concern for the globe, rather than for a region. It is a mark of progress, of a kind. Since 1970, there has been a series of near-apocalyptic warnings, not just from the 'green' movements, but from national political leaders and from senior scientists and economists about the dangers of uncontrolled economic growth, of the population explosion, of the alarming disparity between the rich and poor nations, and the wasteful exploitation of resources. Since 1988, these warnings have become more urgent, and more apocalyptic. In the winter of 1989/90, alarm about the environment became, for a time, smothered by an extraordinary series of political upheavals within the Communist blocs — in China, the Soviet Union, Poland, Czechoslovakia, Germany, Bulgaria and Romania, and even in Outer Mongolia; events which appeared to signal the end of the Cold War which has warped human progress for more than four decades. People started talking about the end of history, and also the end of nature. Neither history, nor nature, of course, will come to an end. But both are likely to be very different in the decades ahead. There is supposed to be an old Chinese curse: 'May you live in interesting times'. The interesting times are just about to begin.

Index

acid rain 23, 77-80 (volcanic)
    123, 131
aerosol propellants 98, 99
Aids 169, 171, 109, 187, 188,
    215-6
Albedo effect 39, 73, 113, 142,
    179, 205
algae 40
Alvarez, Luis and Walter 129
Amazon 46, 176, 177, 179-182
    economic value 184-7
amino acids 21
Antarctic 24, 28, 39, 101-3, 142
Arctic 39, 40, 103-4, 142
arms race 211-3, 218, 219
Arrhenius, Svante August 27
arthropods 166-7
asteroids 129-33
atmosphere (mass and
    composition) 19
Azadirachta indica 189-90

bacteria 22, 163, 164
Balfour, Arthur 195
Baltic 85
Bangladesh 69-71

calcium carbonate 41, 56
calendar 136-7
Callendar, G.S. 64

camels 208
Cameroon 117-8
Canada 72, 142
carbon 35
    carbon cycle 35, 37, 38, 46,
        47
    fixed by phytoplankton 111
    soil content 49
carbon dioxide
    content of atmosphere 23
    fluctuations 24, 25
    from limestone 56
    heat trap 28
    increase 59, 60
    Lake Nyos 118
    ocean disposal 38, 41, 143
    surplus 35
    removed by life 147
carbon monoxide 23, 77, 80
carcinoma 108
Carson, Rachel 83
cataracts 110
catastrophism 123
cement (and carbon dioxide)
    56-7
China 66, 67, 73, 106, 168, 217
chlorine 97, 100, 102
chlorofluorocarbons (CFCs) 23,
    30, 51, 87-114
Churchill, Winston 195
coastlines 66
comet 129-34
continental drift 133, 142-3
coral reefs 68

crops
  carbon takeup 44
  in drought 73-4
  under threat 194-200
cyanobacteria 112

Darwin, Charles 68, 128
DDT 82-3
deltas 49, 50
deserts 201-10
Dickens, Charles 122
dimethyl sulphide 37, 95
dinosaurs 28, 32, 125-8, 132
DNA 107
droughts 28, 46, 66, 69, 203-6

Ecuador 172
Egypt 69, 176
energy
  conservation 61
  cycle 32, 33
  growth 58
  sources 57-8
energy budgets (hummingbird,
      elephant, whale, human)
      53-5
erosion 45, 49, 50, 51, 201-10
extinction 123-4, 127-8, 131-5,
      162-3, 167-72
extreme events (droughts,
      floods, storms) 28, 66, 70-2

Farman, Joe 101
feedback 30, 43, 44, 46, 49, 52,
      61
Fermi, Enrico 118-19
fishery harvest 112
floodplain 67, 69
Florida dusky sparrow 168
forest
  acid rain 80
  destruction 46, 169
  fires 131, 152
  sustainable harvests 184-93
  tropical 172, 174-93
fossil fuels 34

Gabon 119
gaia 33, 94
gene banks 199-200
Genesis, book of 15, 25
genetic bases 194-200
glaciers 29
global average temperatures 25,
      28, 60, 64
greenhouse gases 24
greenland 24, 28, 43, 141, 144,
      145
gulf stream 39
Guyana 68

Halley Bay 101
Harrieson, Newton 211-12
Hilo, Hawaii 120
Hiroshima 149
Holland 67
hydrogen peroxide 37
hydrogen sulphide 117

ice ages 25, 43, 48, 60, 66, 114,
      135, 143-5, 147, 155
ice caps 29, 30, 37, 39
immunization 214-15
infant mortality 215
infra-red radiation 24, 28, 90
insect pests 48
ionosphere 92
iridium 129, 131

keystone species 177-79
Kiribati 68
Kiwi 168

laterite 182
lead 77, 81, 87-9
life expectancy 216
life's origins 21
limestone (conversion) 56
Lovelock, James 33

maize 191, 194-5, 197-9
Maldive atolls 68
Mariner 9, 146
Marland, Gregg 45
Mars 21, 63, 90, 93, 101, 130,
      146-9

Mediterranean 69, 108
melanin 108
melanomas 100, 108-9
Mesozoic 28
methane 30, 37, 51, 52, 97
Mexico 194
Mexico City 78, 217
Miami 67
Midgley, Thomas 87, 88, 95
military spending 215, 218-9
Miller, Stanley 21
Mississippi 69
Molina, Mario 99
Mt St Helens 164
Montreal Protocol 101, 104-6
Myers, Norman 155, 171

Nagasaki 130
Nasa 99, 101
Nemesis 134
New England 122
New Zealand 81, 167-8
Nile 69
nitric oxide 97, 98, 102
nitrogen dioxide 97, 98
nitrogen fertilizers 51
nitrous oxide 30
notornis 168
nuclear testing 98
nuclear winter 63, 131, 140-57
Nyos, Lake 117-18

ocean warming 38
    science 42
Oklo 119
oxygen
    and life 21, 22
    and the ozone layer 89-94
    demand 55
ozone 30, 86-114

Panama 165-66, 176
panspermia 27
Permian 124, 127
permafrost 34
PCBs 83, 85
photosynthesis 21, 39, 40, 45,
    91, 107, 147

phytoplankton 37, 41, 45, 111,
    143
Poland 78
pollution (industrial) 77-80,
    201-2
population growth 51, 59,
    213-19

Rhone 67
Rome 80
rosy periwinkle 187
Rotorua 81
Rowlands, Sherwood 99
rubber 190-91

Sagan, Carl 148-49, 153, 211
salt tolerant plants 209
St Paul's 80
San Andreas fault 120
sea levels 29, 30, 49-50, 66-70
Shanghai 67
Siberian tiger 170
solar cycles 136-40
species 163-67
Steinbeck, John 66
storms 28, 49
stratosphere 93, 98, 121, 131,
    148, 152-3
sulphur dioxide 23, 77-80, 97
sun
    and the ozone layer 90-114
    solar radiation 24
sunspots 141

Taj Mahal 80
Tambora, Mt 121
temperature changes 25, 180
Tesla, Nikola 211
thermonuclear war 145, 149-57
time measurement 136-38
Tolba, Mostafa 104
topsoil 206
toxic metals 77, 131
trees (carbon takeup) 44, 45, 46
tsunami 120, 130
tundra 48, 52
Turner, J.M.W. 122

ultraviolet (and ozone layer)
    90-4, 100, 102, 104, 107,
    110-13, 132, 155
UN Environment Programme
    (UNEP) 100, 104, 106, 209
uranium 118-19

Venus 21, 28, 62, 146-48
Vienna Convention 101
volcanic gases 21

eruptions 98, 121-23, 164

water cycle (and water
    properties) 34
Weizmann, Chaim 195
Wigley, Tom 30

Yangtse 67

Zurich 122